155
Advances in Polymer Science

Springer

Berlin
Heidelberg
New York
Barcelona
Hong Kong
London
Milan
Paris
Singapore
Tokyo

New Polymerization Techniques and Synthetic Methodologies

With contributions by
M. Biswas, I. Capek, C.-S. Chern, D. Mathew,
C.P. Reghunadhan Nair, K.N. Ninan, S. Sinha Ray

 Springer

This series presents critical reviews of the present and future trends in polymer and biopolymer science including chemistry, physical chemistry, physics and materials science. It is addressed to all scientists at universities and in industry who wish to keep abreast of advances in the topics covered.

As a rule, contributions are specially commissioned. The editors and publishers will, however, always be pleased to receive suggestions and supplementary information. Papers are accepted for „Advances in Polymer Science" in English.

In references Advances in Polymer Science is abbreviated Adv. Polym. Sci. and is cited as a journal.

Springer APS home page: http://link.springer.de/series/aps/ or
http://link.springer-ny.com/series/aps
Springer-Verlag home page: http://www.springer.de

ISSN 0065-3195
ISBN 3-540-41435-5
Springer-Verlag Berlin Heidelberg New York

Library of Congress Catalog Card Number 61642

Springer-Verlag Berlin Heidelberg New York
a member of BertelsmannSpringer Science+Business Media GmbH

© Springer-Verlag Berlin Heidelberg 2001
Printed in Germany

Typesetting: Data conversion by MEDIO, Berlin
Cover: MEDIO, Berlin
Printed on acid-free paper   SPIN: 10706242   02/3020hu - 5 4 3 2 1 0 *ac*

4-10-01
ac

√CD
4-17-01

Editorial Board

Advances in Polymer Science
Now Also Available Electronically

For all customers with a standing order for Advances in Polymer Science we offer the electronic form via LINK free of charge. Please contact your librarian who can receive a password for free access to the full articles. By registration at:

http://link.springer.de/series/aps/reg_form.htm

If you do not have a standing order you can nevertheless browse through the table of contents of the volumes and the abstracts of each article at:

http://link.springer.de/series/aps/

There you will find also information about the

- Editorial Bord
- Aims and Scope
- Instructions for Authors

Contents

Cyanate Ester Resins, Recent Developments

C.P. Reghunadhan Nair[1], Dona Mathew, K.N. Ninan

Propellant and Special Chemicals Group, Vikram Sarabhai Space Centre
Trivandrum-695022, India
[1] e-mail: cprnair@eth.net

The search for advanced, high performance, high temperature resistant polymers is on the rise in view of the growing demand for polymer matrix composites that are to meet stringent functional requirements for use in the rapidly evolving high-tech area of aerospace. Cyanate esters (CEs) form a family of new generation thermosetting resins whose performance characteristics make them attractive competitors to many current commercial polymer materials for such applications. The chemistry and technology of CEs are relatively new and continue to evolve and enthuse researchers. The CEs are gifted with many attractive physical, electrical, thermal, and processing properties required of an ideal matrix resin. These properties are further tunable through backbone structure and by blending with other polymer systems. The structure-property correlation is quite well established. Several new monomers have been reported while some are commercially available. The synthesis of new monomers has come to a stage of stagnation and the present attention is on evolving new formulations and processing techniques. The blends with epoxy and bismaleimide have attracted a lot of research attention and achieved commercial success. While the latter is now known to form an IPN, the reaction mechanism with epoxy is still intriguing. Extensive research in blending with conventional and high performance thermoplastics has led to the generation of key information on morphological features and toughening mechanisms, to the extent that even simulation of morphology and property has now become possible. Despite the fact that the resin and its technology are nearly two decades old, the fundamental aspects related to curing, cure kinetics, reaction modeling, etc. still evince immense research interest and new hypotheses continue to emerge. The system enjoys unprecedented success for applications in microelectronics, aerospace, and related areas. CEs tends to be the resin of choice in advanced composites for aerospace and high-speed electronics. The objective of the present article is to analyze recent developments in the chemistry, technology, and applications of cyanate esters. It mainly focuses on the advancements in research and development related to fundamental and applied aspects of cyanate esters during the last few years. It also gives a brief account of the overall scenario of the developments in this area, prior to discussing recent trends in detail.

Keywords. Cyanate ester, Polycyanurates, Phenolic-triazine, Polymer matrix composite, Cure kinetics, Polymer blends, Cyanate-epoxy blend, Bismaleimide-triazine resins, Aerospace structure

Advances in Polymer Science, Vol. 155
© Springer-Verlag Berlin Heidelberg 2001

List of Abbreviations

AcAc	Acetyl acetonate
AFM	Atomic force microscopy
BACP	2,2-bis(3-allyl-4-cyanatophenyl) propane
BACY	2,2-bis(4-cyanatophenyl) propane (bisphenol A dicyanate)
BECY	1,1-bis(4-cyanato phenyl) ethane
BFCY	2,2-bis(4-cyanatophenyl)1,3-hexaflouro propane
BMI	Bismaleimide
BMIP	2,2-bis[4-(4-maleimido phenoxy) phenyl] propane
BMM	Bis(4-maleimido phenyl) methane
BNM	Butyl acrylate-Acrylonitrile-HPM copolymer
BPCP	2,2-Bis(3-propenyl-4-cyanatophenyl) propane
BRTM	Bleed resin transfer molding
BT	Bismaleimide-triazine resins
CE	Cyanate ester
CEPOX	BACY-EPN 1139 blends
CME	Coefficient of moisture expansion
CMP	Compimide
Cp-	Cyclopentadienyl group
CTE	Coefficient of thermal expansion
DBTDL	Dibutyl tin dilaurate
DDS	Diaminodiphenylsulfone
DGEBA	Diglycidyl ether of bisphenol A
D_k	Dielectric constant
D_f	Dissipation factor
DMA	Dynamic mechanical analysis
DMAC	Dimethylacetamide
DSC	Differential scanning calorimetry
E	Activation energy

EPN 1139 Low molar mass novolac epoxy resin
ETT   Effective time theory
E' Storage modulus (flexural)
E'' Loss modulus (flexural)
FTIR   Fourier transform infrared spectroscopy
G_{1C} Critical strain energy release rate (mode I fracture)
G_{1IC} Critical strain energy release rate (mode II fracture)
HPLC   High performance liquid chromatography
HPM N-(4-hydroxy phenyl) maleimide
ILSS   Interlaminar shear strength
ILT   Interlaminar toughening
IPN   Interpenetrating polymer network
K_{IC} Fracture toughness (mode 1)
LC   Liquid crystal
LIPN   Linked interpenetrating polymer network
MEK   Methyl ethyl ketone
MIES   Metastable impact electron spectroscopy
MMA   Methyl methacrylate
NMR   Nuclear magnetic resonance spectroscopy
PCB   Printed Circuit Board
P-T   Phenolic triazine
PES   Poly ether sulfone
PET   Poly(ethylene terephthalate)
Phr   Parts per hundred part
PMC   Polymer matrix composite
PMR   Polymerizable monomeric reagents (LARC)
RT   Room temperature
RTM   Resin transfer molding
SBSS   Short beam shear strength
SEC   Size exclusion chromatography
SIPN   Simultaneous interpenetrating polymer network
SPM   Styrene-HPM-phenyl maleimide copolymer
TBA   Torsional braid analysis
T_g Glass transition temperature
TGA   Thermogravimetric analysis
T_{gel} Gel time
TMA   Thermo mechanical analysis
TGMDA   Tetraglycidyl methylene dianiline
Tol-   Tolyl group
TsOH   p-toluene sulfonic acid
TTSP   Time temperature superposition principle
UD com-
posite   Uni-directional composite
UPS   UV photoelectron spectroscopy
XPS   X-ray photoelectron spectroscopy

1
Introduction

Advanced thermosetting polymers find an undisputed place in the design of polymer matrix composites (PMC). High strength, lightweight, composite materials have helped transform many creative ideas into technical realities. The application of these materials has helped boost many industries, including transportation, communication, construction and leisure goods. An area that has immensely benefited from the boom in PMC technology is aerospace where the impact of these materials can be seen in the space program, rocket motors, stealth military aircraft, commercial transport, and so forth. This recognition is attributable not only to the economy of weight-saving but also to the key properties of PMCs, like high specific stiffness and strength, and the ability for generating near zero coefficient of thermal expansion and amenability of the systems for tailoring the physical and mechanical properties. Weight reduction in air/space craft and launch vehicles can obviously lead to benefits like improved performance and fuel economy, and increased range and payload capacity. PMCs constitute several vital components of satellites and launch vehicles ranging from motor cases and nozzle components in rocketry to systems like honeycomb structures, equipment panels, cylindrical support structures, pressurant bottles, solar array substrates, antennas, etc. in satellites. Recent reports show that in modern launch vehicles meant for small satellites, the structure can be composed of PMCs to the extent of even greater than 80%. This proliferation of PMCs is bound to show an upward trend in days to come. Examples of PMC components and systems are numerous. To cite a few, rocket motor cases of the space shuttle's solid boosters comprise 30 tons of graphite-reinforced epoxy composites. Antenna reflectors of NASA's advanced communication technology satellite are made of sandwich structures of a honeycomb core, bonded to graphite/epoxy skin. PMCs have generally three-fold higher specific strength than conventional aerospace materials such as titanium and their use in spacecraft would result in decreased system weight and enhanced payloads.

1.1
State-of-the-Art Matrix Systems for Advanced Composites

1.1.1
Epoxies

Although most of the benefits of PMCs derive from the excellent properties of high strength reinforcing fibers, the matrix plays an equally important role in defining these advantages and dictating many of their characteristics. Composites for specific applications with desired properties are realizable through correct choice of the resin, reinforcement, and their perfect processing. The bulk of the aerospace composite materials centers on epoxies in combination with graphite/glass and to a limited extent, with organic reinforcements like polyaramid. Their attractive features include ease of processing and handling convenience, good mechanical properties due to excellent adhesion to a variety of reinforcements, toughness, and low cost. Depending upon the end application, different varieties of epoxy resins are in use [1]. All resin formulations contain many additives, fillers, viscosity modifiers, flame retardants, coupling agents, cure accelerators, etc. The curing agents are selected among amines, phenols, thiols, anhydrides, carboxylic acids [2]. The inherent brittleness, a source of microcracking, has been a major concern limiting the use of epoxy-based composites in areas such as wing, tail, and fuselage structures. Several tough epoxy formulations based on blends with elastomers, and thermoplastics such as polyether sulfones, polyetherimides, and polyhydantoins have been introduced. In the majority of cases the resin is not available as such except as prepregs. Many earlier aerospace applications such as helicopter rotor blades, floorboards, fairings, etc., required low T_g curatives and for more demanding structures, 177 °C curing systems such as those based on TGMDA and DDS are needed. In aircraft, their use is limited to secondary non-load-bearing structures such as engine nascelles, cowlings, fan containment cases, flaps, ailerons, spoilers, rudders, elevators, landing gear doors, etc. [3]. This figure is, however, increasing, particularly in the US commercial aircraft industry [4]. Epoxy resin matrices still continue to dominate the structural composites in aerospace applications. Lightweight motor cases use combinations of polyaramid and epoxy systems. Satellite support structures, antennas, yoke and solar panel substrates are preferably made of epoxy-carbon composites. Advanced tetraglycidyl resins can cater for many critical requirements of composites for satellite applications. In addition to high performance aerospace composites, they are used in coatings, adhesives, flooring, and electrical applications. The nature and amount of curing agents and their concentrations are critical in deciding the properties of the cured network [5, 6].

However, even with advanced epoxy composite systems, the upper use temperature is limited to 150–80 °C although in very selective cases it has been pushed to the 200 °C barrier. Advancements in aerospace technology warrant pushing up the upper use temperature of PMCs and, consequently, a need has arisen for alternate systems with higher service temperatures. Such thermally stable materials are needed for aircraft engine parts and skin of supersonic aircraft and missiles. In launch vehicles, the composite motor cases would preferably be made from high temperature resistant matrices, so that the thickness for thermal insulation can be reduced, leading to improvements in mass-ratio and thrust. In satellites, the thermally stable composite-structural components ensure better dimensional stability. In such situations, epoxy resins, despite all their good attributes, cannot meet these challenges and need to be replaced by a resin of superior thermal and hygrothermal characteristics. Epoxy resins are also not generally recommended for application in severe conditions due to poor hot/wet performance and high moisture sensitivity [7]. For electrical/electronic applications, an alternate resin system with better dielectric properties and reduced moisture absorptivity will be preferred.

1.1.2
Polyimides

Polyimides were in the limelight as replacement for epoxies, attracting the aerospace market for sometime. Polyimides, with their service temperatures ranging from 200 °C to 280 °C offer improved high temperature properties compared to epoxy systems. A number of polyimide systems that are well suited for use as matrix resins, adhesives, and coatings for high performance applications in the aerospace and electronics industries have been developed [8–11]. Condensation polyimides are formed by a two-step process involving formation of a polyamic acid intermediate followed by its imidization. Thus, the processing times are much longer and require the application of high pressure for consolidation, due to evolution of water or alcohol as condensation by-products. Major disadvantages of these systems also include poor shelf-life and processing difficulties [12]. The growing demand for polyimides that are processable, solvent resistant, and with good thermal stability provided a strong incentive for developing processable polyimides. One strategy is to develop melt processable or soluble thermoplastic condensation polyimides in the state of full imidization [13–15]. The structural features conducive to this include preformed polyimides containing aliphatic and other flexible spacers, bridging functional groups, polyimides with bulky substituents, etc. The flexible spacers comprise groups such as perfluoroalkyl, alkyl ether, disilyl, siloxane, etc., and typical flexible bridge groups are ether, carbonyl, silyl, sulfone, hexafluoro isopropyl, phenyl phosphonyl, etc. These groups can originate either from the diamine, or from the dianhydride, or from both. Accordingly, a wide spectrum of properties can be imparted to the resultant thermoplastic polyimides. Condensation polyimides may be rendered tractable by way of substitution of bulky groups on the polyimide backbone too. Such groups include phenyl, tertiary butyl, fluoroalkyl, cardo, diphenyl thiophene, phthalimide, etc. However, due to their thermoplastic nature, the success of these materials as matrices in high performance composites is questionable.

The matrix system encompassing both processability and good thermo-mechanical profile are the addition-type polyimide systems. These are low molar mass, multifunctional monomers or prepolymers that carry additional curable terminal functions and imide functions in their backbone [16–19]. The bismaleimides, bisitaconimides, and bisnadimides are the versatile ones in this category. Of these, bimaleimides still dominate the high temperature polymer scenario for aerospace applications. But these polymers are extremely brittle. Tough compositions need blending with elastomers and/or chain extension using diamines. A recent development for tough BMI compositions is based on the blends with diamines, diallyl phenols, and with other bismaleimides. [20–22]. Commercial addition-cure formulations, based on co-reaction of bisallyl phenols and bismaleimides such as the Matrimide-5292 of Ciba-Geigy, are available. This is one of the leading matrix resins for carbon fiber composites for advanced aerospace applications. Several modified versions of BMI, particularly the PMR-type resins, are currently in use in the aerospace industry [23]. Variations of the PMR type addition polyimides with nadic [23], paracyclophane [24], acetylene [25], and cyclobutene [26] terminals have also been reported. In combination with graphite, many of these modified imide systems have been used in aircraft engine components in place of titanium with good gains in weight saving. However, the PMR systems also possess inherent difficulty in processing. They need a solvent medium for impregnation and the curing (through a condensation reaction) requires high temperature. Release of volatiles during condensation cure of composites is not very desirable as it can result in blistering and void formation, leading to inferior laminates. To obviate these problems in part, ethynyl terminated imide and isoimide and norbornene terminated imide (PMR-15) oligomers were introduced for high temperature applications [27]. However, these systems too are infamous for their brittle nature and offer only moderate improvement in processability over PMR type resins.

1.1.3
Phenolic Resins

Although phenolics cannot be substitutes for epoxies and polyimides, their composites still find a major market in non-structural components, particularly in thermo-structural application in the aerospace industry. Good heat and flame resistance, ablative characteristics, and low cost are the hallmarks of phenolics which find numerous applications including aircraft interiors, automotive components, rocket nozzles, appliance moldings, etc. [28–30]. However brittleness, poor shelf-life, and need for high pressure curing are the major shortcomings of phenolic resins in processing and for structural composite applications. Blending with elastomers, which risks the thermal properties, is a means to obviate brittleness. Research attention is now focused on addition-curable phenolics, which are alternatives to condensation type phenolics, avoiding the need for high-pressure cure [31–33]. However, most of them require high cure temperature, posing processing difficulties.

1.1.4
Cyanate Ester Resins

Against these backdrops, cyanate ester emerged as a compromise system. This new generation thermoset resin encompasses the processability of epoxy resins, thermal characteristics of bismaleimides, and the heat and fire resistance of phenolic resins. The highly desirable properties like built-in toughness, good dielectric properties, radar transparency, and low moisture absorption make them the resin of choice in high performance structural aerospace applications and in the electronics industry. A comparison of conventional high performance thermoset matrix resins with cyanate esters is given in Table 1. Figure 1 shows T_g vs strain capabilities of various matrix systems. The relative position of the various systems indicates their comparative toughness and temperature endurance ca-

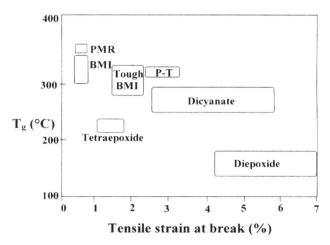

Fig. 1. Service temperature and toughness of various thermoset matrix polymer systems

Table 1. Comparison of common high performance thermosetting polymer matrices

Property	Epoxy	Phenolic	Toughened BMI	Cyanate ester
Density (kg.m^{-3})×10^{-3}	1.2–.25	1.24–.32	1.2–.3	1.1–.35
Use temperature (°C)	RT-180	200–250	~200	~200 (250)[a]
Tensile modulus (MPa)×10^{-3}	3.1–3.8	3–5	3.4–4.1	3.1–3.4
Dielectric constant (1 MHz)	3.8–4.5	4.3–5.4	3.4–3.7	2.7–3.2
Cure temperature (°C)	RT-180	150–90	220–300	180–250
Mold shrinkage (mm mm$^-$)	0.0006	0.002	0.007	0.004
TGA onset (°C)	260–340	300–360	360–400	400–420

[a] P-T resin

pabilities. The positions for PMR and PT resins are judged by extrapolation of their composite properties and are only approximate. The phenolic resins have not been positioned in the chart, as they do not possess a defined T_g regime. The fully cured, brittle phenolic resin usually exhibit elongation of the order of 2–3% and starts decomposing prior to glass transition. The dicyanates can be found to occupy a medium position.

Reports on the use of cyanate ester resins in aerospace structures and related applications are scanty. Practically all the literature pertaining to their trials for potential aerospace applications are in the form of patents. All indications are that this system with its many attractive features is poised to replace the epoxies in composites for specific applications. Presently, the bulk of cyanate ester resins is consumed by the high-speed electronics industry, which was once dominated by epoxy resins. The emerging importance of cyanate systems can be understood from the increasing research interest in this area, evident in terms of a large number of publications and patents. In view of its undisputed superiority over conventional matrices and the assured place as the unique supreme resin of the immediate future, it is of great significance to keep watch on developments in this area. The objective of the present article is to analyze recent development in the chemistry and technology and applications of cyanate esters. There have been a few review articles devoted to various aspects of cyanate esters, which are referred to in this text at appropriate places. The present article mainly focuses on research and development related to fundamental and applied aspects of cyanate esters over the last four to five years. It also gives a brief account of the overall scenario of the developments in this area, prior to discussing recent trends in detail.

2
Synthesis and Reactions of Cyanate Esters

2.1
Synthesis of Cyanate Esters

Cyanate esters (CEs) are formed in excellent yields by the reaction of the corresponding phenols with cyanogen halide [34]. The reaction scheme is shown in Scheme 1a. The reaction is usually carried out in solution, in the presence of a tertiary amine as the acid scavenger at very low temperatures. Since the trimerization reactions of cyanate esters are highly prone to catalysis by spurious impurities, the most difficult aspect of cyanate ester synthesis is their scrupulous purification. Low molar mass esters are purified by distillation or recrystallization. Polymeric cyanates are purified by repeated precipitation in non-solvents such as water, isopropanol, etc. While distillation and recrystallization lead to pure materials, the precipitation method for polymeric cyanates is not always conducive to obtaining pure materials. A recent patent application claims a purification method for polymeric cyanates based on treatment with cation and anion exchangers [35].

Researchers have conducted a number of studies highlighting the synthesis of CEs, reaction conditions including solvents and catalysis [36–38], etc., but most of the details are covered by patents. The solvents selected include acetone [39], methyl isobutyl ketone [40], dichloromethane [41], etc., and acid scavengers chosen mainly include triethyl amine and pyridine [39, 41]. The use of trialkylamines often leads to the formation of dialkyl cyanamides as side products, which are known to catalyze the reactions of cyanate esters, reducing the yield. The formation of these impurities depends on the mode and rate of addition of reagents, reaction temperature, etc. Factors like stirring rate, dropping rate, and

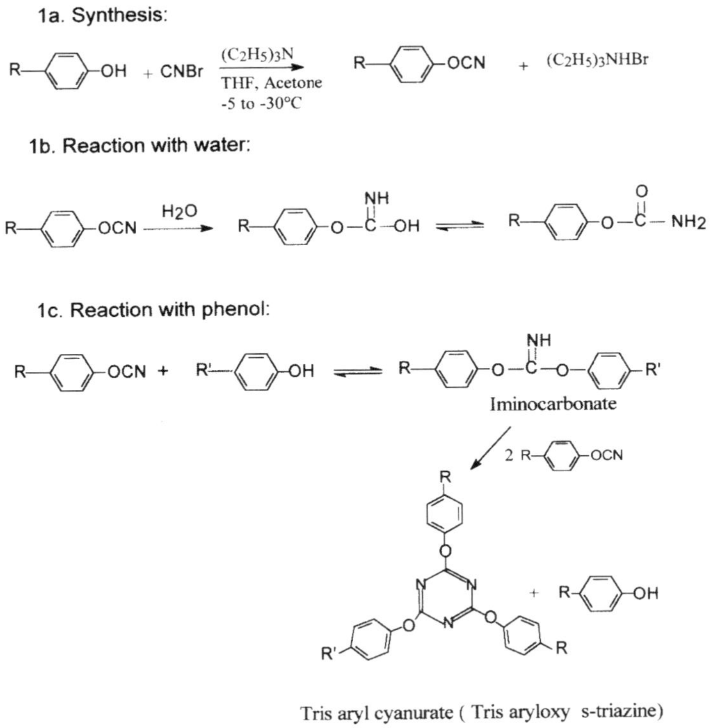

1a. Synthesis:

1b. Reaction with water:

1c. Reaction with phenol:

Tris aryl cyanurate (Tris aryloxy s-triazine)

Scheme 1. Synthesis of cyanate ester and its reaction with water and phenol

reaction temperature have been shown to affect the reaction to a large extent [42]. The synthesis is preferably performed under moisture-free, optimized conditions, usually in an inert atmosphere. Another method for obtaining aryl cyanate is by the thermolysis of phenoxy 1,2,3,4-thiotriazole [43] which, for the moment, is only of academic interest.

2.2
Reactions of Cyanate Esters

The carbon atom of the -OCN group is strongly electrophilic and hence is highly susceptible for attack by nucleophilic reagents. As a result, cyanate groups undergo a variety of reactions, e.g., with water, phenols, and other hydrogen-donating impurities, transition metal complexes, imidazole, etc. With water, it produces carbamates [44] as shown in Scheme 1b. These products, in turn, can act as catalysts for the thermal curing of cyanate groups through cyclotrimerization to rigid, heterocyclic phenolic triazine networks. Phenols also react with aromatic cyanates in the presence of a base to give bisaryl iminocarbonates, which generally undergo reversible dissociation with the liberation of more acidic phenol. Reaction of this intermediate with cyanate ester leads to formation of cyanurate (or *s*-triazine) derivatives as shown in Scheme 1c. In this process too, the more acidic phenol is liberated and the catalysis of cyanate trimerization by phenols is presumed to proceed this way.

CEs are known to react with groups like epoxy, bismaleimide, etc. These reactions are described in detail in Sect. 8 devoted to blends. Although reaction with triple bonded compounds has been postulated, evidences for this remain to be furnished.

3
Structure-Property Relations

3.1
Dicyanates

A variety of CEs with tailorable physico-chemical and thermo-mechanical properties have been synthesized by appropriate selection of the precursor phenol [39, 40]. The physical characteristics like melting point and processing window, dielectric characteristics, environmental stability, and thermo-mechanical characteristics largely depend on the backbone structure. Several cyanate ester systems bearing elements such as P, S, F, Br, etc. have been reported [39–41, 45–47]. Mainly three approaches can be seen. While dicyanate esters are based on simple diphenols, cyanate telechelics are derived from phenol telechelic polymers whose basic properties are dictated by the backbone structure. The terminal cyanate groups serve as crosslinking sites. The polycyanate esters are obtained by cyanation of polyhydric polymers which, in turn, are synthesized by suitable synthesis protocols. Thus, in addition to the bisphenol-based CEs, other types like cyanate esters of novolacs [37, 48], polystyrene [49], resorcinol [36], *tert*-butyl, and cyano substituted phenols [50], poly cyanate esters with hydrophobic cycloaliphatic backbone [51], and allyl-functionalized cyanate esters [52] have been reported.

In recent work on new monomers, aromatic dicyanate esters with various bridging groups such as phenyl phosphine oxide (see Scheme 2), sulfone, and carbonyl were designed and their cure and thermal characteristics correlated to structure. These groups increased the reactivity and eliminated the need for cure catalysts. The arylene ether phenyl phosphine displayed several attractive features like low softening point, wide processing window, high T_g, and higher char, suggesting potential fire retardancy. The thermal stability of these cyanate ester polymers exceeds those of multifunctional epoxies and compares favorably with that of bismaleimides. Although the authors have attributed the low cure temperature to the electron-withdrawing effect of the bridge groups, it is quite likely that the cure is initiated at low temperature due to catalysis by adventitious impurities adsorbed on the hetero atom-containing backbone [39]. Similarly, bisphenol-based aryl dicyanate esters containing phenyl phosphine oxide moieties were found to exhibit good flame retardancy and oxygen plasma resistance

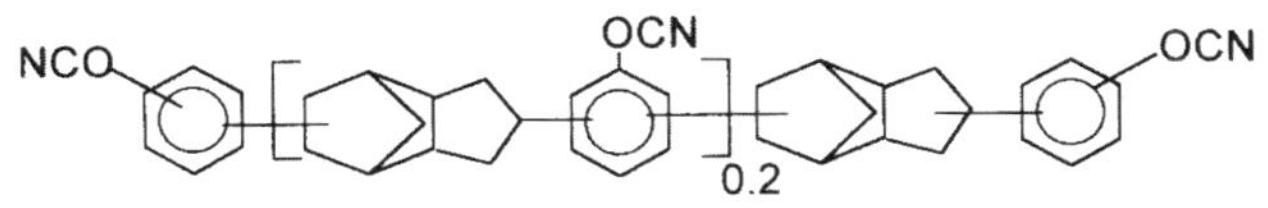

Cyanate ester - XU 71787.02L

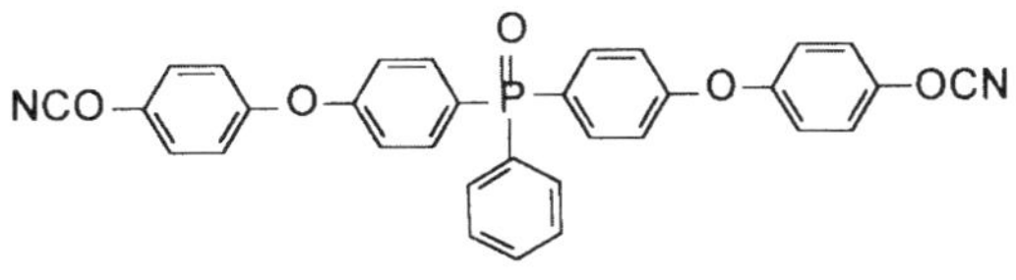

Bis [4-(cyanatophenoxy) phenyl] phenyl phosphine oxide

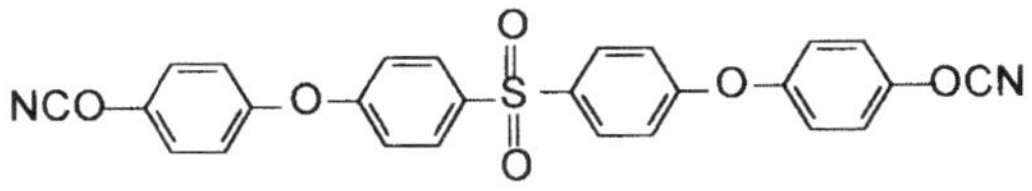

Bis 4-(cyanatophenoxy) phenyl sulfone

Scheme 2. Structure of some new cyanate esters

[53]. The latter property is desirable for its application in low-earth orbit satellite structural components. The fire resistant materials currently used in commercial aircraft cabins do not meet the goal of generating survivable aircraft cabin conditions for 10–5 min in post-crash fuel fire [54]. However, the new phosphorus-containing CE resins give some hope and are claimed to exhibit good flame resistance and find use in the aircraft interior applications [55]. Similar phosphorus-containing cyanate esters have been synthesized and evaluated for their flame resistance [56]. Although they exhibited UL V-0 flame resistance, the heat release rates of these materials were higher than desired. They also generated a significant amount of smoke during combustion [57]. These problems were partly solved by synthesis of a new series of cyanate esters (of undeclared structure) with extremely low heat release rate and higher char yield [58]. An alternative approach for realizing toughened, fire resistant, and cost-effective cyanate ester resin is by way of cyanate ester layered silicate nano composite. Although silicate-polymer nanocomposites are quite well reported [59], such nanocomposites based on cyanate ester is new. Dispersion at nanometer levels of layered silicate, montmorillonite in novolac cyanate ester resin improved their flame resistance [60]. Melamine treatment of the montmorillonite yielded exfoliated montmorillonite in the cured cyanate nano composite. This reduced the peak heat release rate by 50%. The reaction of the amine on melamine with the cyanate to form isourea and the ionic bond of melamine with the silicate is expect-

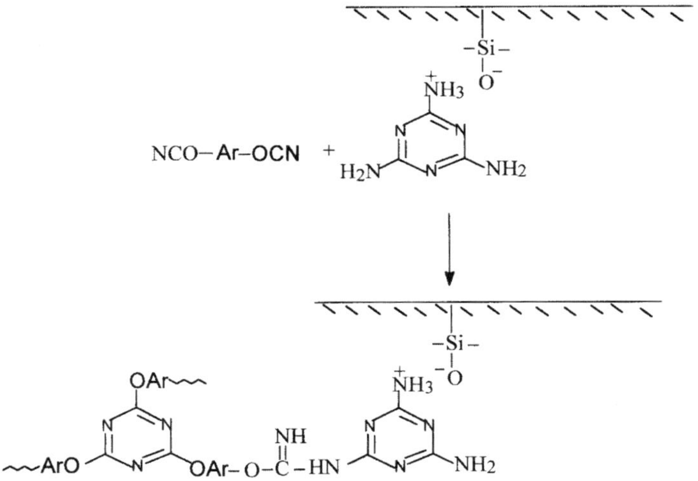

Scheme 3. Tethering of P-T network to the silicate layer via reaction with melamine

ed to tether the organic polymer network to the silicate without creating dangling chains in the matrix. This mechanism is depicted in Scheme 3.

Studies have been conducted to investigate the influence of chemical structure of the monomer and crosslink density on ignition temperature, flammability, etc. [61]. Cyanato phenyl substituted cyclotriphosphazenes were synthesized and were thermally cured to phosphazene-triazine cyclomatrix network polymers. The relative ratio of phosphazene and triazine in the network was regulated by the phenol functionality of the cyclophosphazene intermediate. Two polymer networks with phosphazene-triazine ratio 1:1(Pz-Tz-3) and 3:4 (Pz-Tz-4) were synthesized. The idealized structures of the precursors and the phosphazene-triazine polymer (Pz-Tz-3) are shown in Scheme 4. The phosphazene-triazine polymer network exhibited better flame- and thermal resistance in comparison to the pure polycyanurate [62, 63]. The thermal stability and the activation energy for thermal decomposition of the cyclomatrix network polymer increased with crosslink density and the phosphazene content. The comparative thermograms of the phosphazene-triazine polymer and poly(BACY) are shown in Fig. 2. The phosphazene-based polymers left more char residue than poly(BACY) at higher temperature. The flame retardancy measured in terms of the LOI values increased proportional to the phosphazene content and crosslink density. Since flame retardancy showed a linear increase with char content, it was concluded that the flame-retardant action is confined to the condensed phase. This behavior is shown in Fig. 3 [63]. Reports on the synthesis and characterization of bisphenol A monocyanate monoglycidyl ether which can undergo curing to form a tough and strong resin and 2-allyl phenyl cyanate, capable of co-curing with other cyanate groups and olefinic monomers, are also available [64]. These are discussed in Sect. 8. These polyfunctional monomers can act

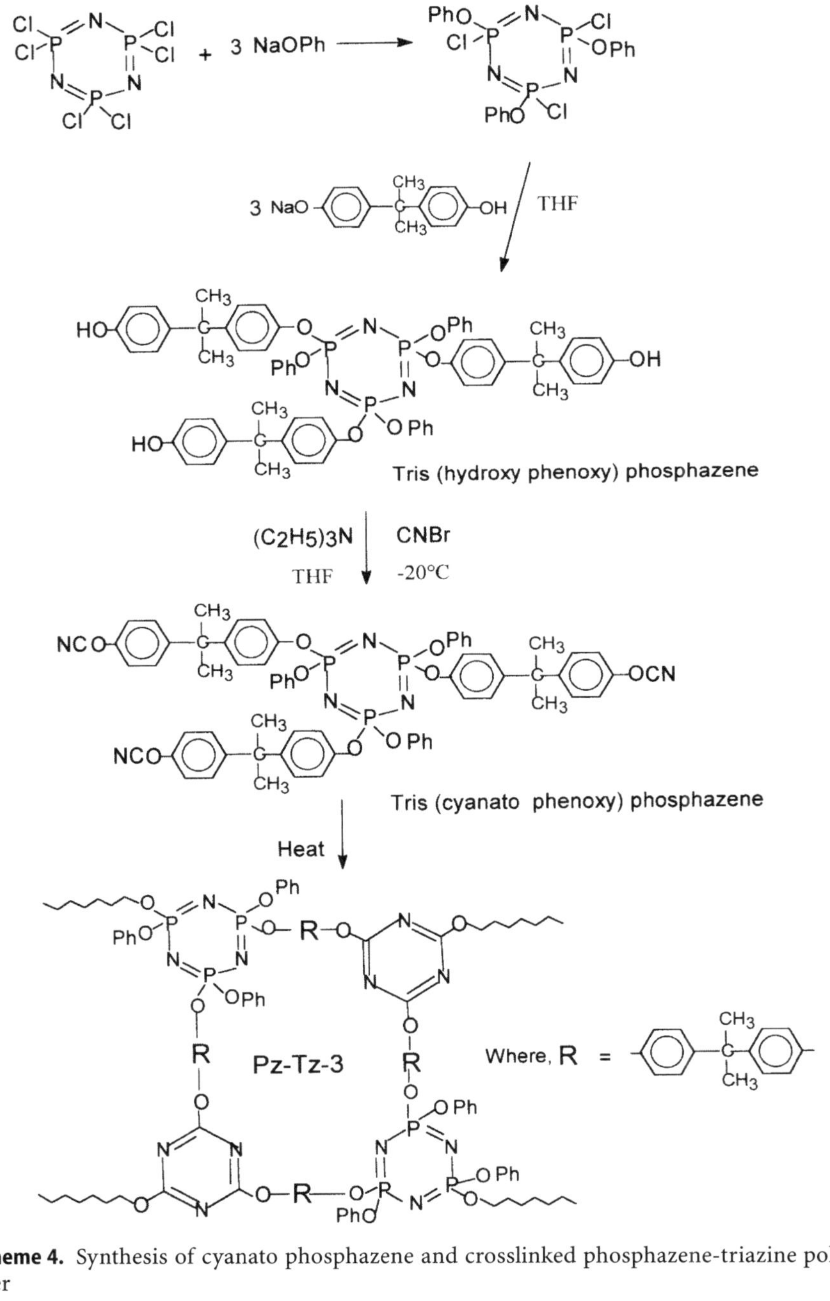

Scheme 4. Synthesis of cyanato phosphazene and crosslinked phosphazene-triazine polymer

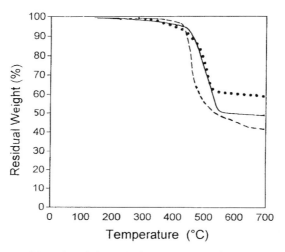

Fig. 2. Thermograms of (- - - -): poly(BACY), (—): Pz-Tz-3 and (•••••••): Pz-Tz-4 in N_2 atmosphere. Heating rate :10 °C/min

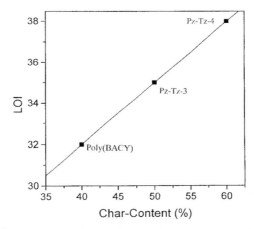

Fig. 3. Relationship between LOI values and anaerobic char residue for phosphazene-triazine polymers

as crosslinking agents for other cyanate resins. In this series, a range of allyl functional cyanate ester oligomers with 4,6-phenylene groups in the backbone were reported by Chaplin et al. [52].

Although a large variety of such experimental cyanate esters have been reported, few have achieved commercial success. Some of these have been listed in Table 2 along with their key properties. The properties are very much dependent on the backbone. Thus, the experimental CE resin XU.71787 developed by Dow

Table 2. Commercial cyanate esters and thei monomer and polymer Properties (*)

Structure of cyanate ester (monomer)	Trade name/Supplier	Melt. pt. (°C)	T_g * (°C)	H_2O absorb* (%)	Dk (1 MHz)
(NCO–C₆H₄–C(CH₃)₂–C₆H₄–OCN)	AroCy B/ Ciba-Geigy BT-2000/Mitsubishi GC Lonsa	79	289	2.5	2.91
(tetramethyl bisphenol dicyanate)	AroCY M/Ciba-Geigy	106	252	1.4	2.75
(NCO–C₆H₄–C(CF₃)₂–C₆H₄–OCN)	AroCy F/Ciba-Geigi	87	270	1.8	2.66
(NCO–C₆H₄–CH(CH₃)–C₆H₄–OCN)	AroCy L/Ciba-Geigi	29	258	2.4	2.98
(structure)	XU-366/Ciba-Geigi RTX-366	68	192	0.7	2.64
(novolac-type structure, n)	Primaset PT/ Lonsa REX-371/Ciba-Geigy	Resinous	> 350	3.8	3.08
(cycloaliphatic structure, 0.2)	XU 71787.02L/ Dow Chemical	Resinous	244	1.4	2.80

* Cured polymer

Chemical Company in 1988, bearing the hydrophobic cycloaliphatic backbone, was claimed to possess good dielectric properties, extremely low moisture absorption, low cure shrinkage, superior thermal cycling tolerance, and retention of mechanical properties at high temperature [47, 51, 65, 66]. Dicyanate esters with mesogenic backbones, exhibiting liquid crystalline properties, have been reported and the discussion related to them can be found under liquid crystalline cyanate esters. Very recently, dicyanate esters with ether and ether-ketone backbones have been described. The T_g of the cured network depended on the length and symmetry of the monomer, T_g being higher for shorter and the para substituted ones. Rigidity of the shorter chain was found to retard the cure reaction. Other monomers exhibited very short processing windows. Unlike dicyanate ester of bisphenol A (BACY) and its homologues, these systems provided higher char residue, to the extent of 60% at 700 °C [67].

3.2
Cyanato Telechelics

Prior to the claim on phenolic triazine by Das et al. [68, 69], thermally stable cyanate telechelics had been reported by Still et al. [70] which, due to the complexity of synthesis merits a special citation. They reported thermally stable biscyanato polymers of phenylated polyphenylenes and polyphenyl quinolines. In these cases, the cyanate groups serve as crosslinking sites, while the basic properties of the backbone could remain practically unaffected. Thermosetting poly(ether ketones) bearing cyanate functions undergo curing to give the triazine derivative with much better mechanical properties than the linear poly(ether ketones). Processing temperature of the former is much lower, thereby providing a broad processing window [71]. The chemical structure of some of these compounds can be found in Scheme 5. Reference to a large variety of cyanate telechelic polymers such as polyether sulfones (PES), polyetherimides (PEI), etc.

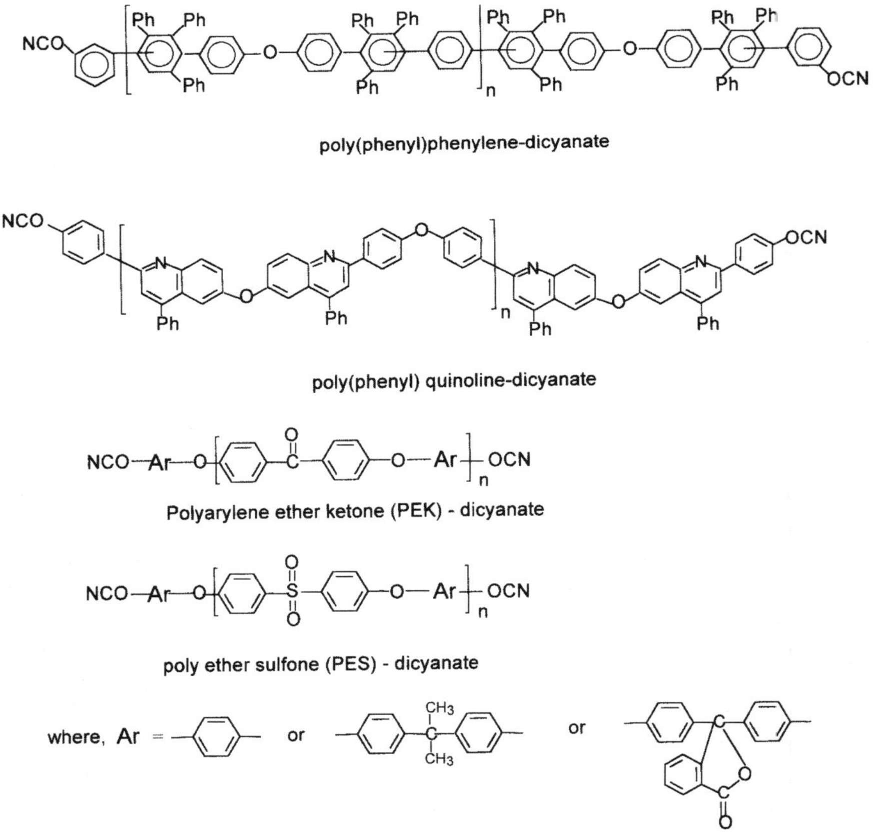

poly(phenyl)phenylene-dicyanate

poly(phenyl) quinoline-dicyanate

Polyarylene ether ketone (PEK) - dicyanate

poly ether sulfone (PES) - dicyanate

Scheme 5. Structure of some cyanate telechelics

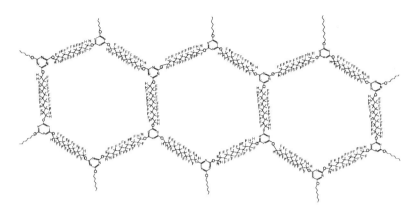

Scheme 6. Fluoro methylene polycyanurate

can be found in the literature, the majority of which are used as matrix modifiers for commercial cyanate esters. Their discussion is therefore more relevant in the section dealing with blends, tougheners, etc. (Sect. 8). Flammability studies on cyanate ester resins, with different backbones and cyanate functionality, have been conducted by Gandhi et al. Heat of reaction as well as thermal stability was dependent on the backbone chemistry of the cyanate esters. Thermal stabilities were indirectly studied by the steady state burning behavior, total heat release, and heat of gasification. Burning behavior was correlated to the crosslink density of the polymers, i.e., with increasing crosslink density; the flammability resistance increased due to formation of thicker char [72].

Although the majority of the publications concerns cyanate esters with aromatic backbone, a few aliphatic cyanate esters are also reported on. Thus, a series of cyanate ester monomers containing fluoromethylene bridges in their backbone has been synthesized and characterized. The expected rearrangement of the alkyl cyanate to the isocyanate is probably precluded by the electron-withdrawing fluoromethylene groups. The oligomers could be represented by the general formula $NCO-CH_2-[CF_2]_n-CH_2-OCN$, where n= 3, 4, 6, or 8. The dielectric constant [2.7–2.3] is the lowest among cyanate esters and moisture absorption [1.67–0.68% at 100 °C] is lower than that of BACY. Their melting points vary from –8 °C to 181 °C depending on the number of fluromethylene bridges in the backbone and glass transition temperature is in the range 84–101 °C, significantly lower than that of aromatic polycyanurates. The low-density polymers have critical surface tension of 40–23 dynes/cm and refractive indices in the range 1.45–1.14. These were achieved without any penalty on thermal stability. Among the series, the 1,6-dicyanatohexa fluoro methylene exhibited optimum properties. The authors proposed a structure for the network comprising macrocyclic rings containing triazines, generating a lot of free volume and thereby contributing to the low dielectric properties. The structure is shown in Scheme 6. However, like many cyanate esters, they needed scrupulous purification to produce predictable cure profile and shelf life [73, 74].

3.3
Cyanate Esters of Poly Phenols and Phenolic Triazines (PT Resins)

Phenolic triazine (PT) resins are formed by the thermal cyclotrimerization of the cyanate ester of novolac [68, 69]. The absence of volatile by-products during cure renders them attractive matrices in void-free composites. PT resins based on novolac phenolic resin possess better thermo-oxidative stability and char yield than their precursors. This is attributed to the fact that the phenolic resins are crosslinked mostly by short methylene bridges. The proximity of the phenolic hydroxyl groups renders these methylene bridges thermo-oxidatively fragile and the degradation process is dependent on the number of dihydroxyphenyl methane groups. It is stated that there is always a thermo-oxidative process during degradation irrespective of the atmosphere. The high oxygen content of phenol is also responsible for this. The degradation process of phenolic resins, proposed by Conley [75] is shown in Scheme 7. PT resins, on the other hand, are crosslinked mostly by triazine phenyl ether linkages, which confer both thermo-oxidative stability and toughness to the system. Structure of a cyanated novolac, and the cured product, phenolic-triazine are given in Scheme 8. The evidence for better thermo-oxidative stability is obtained from the thermal behavior of the systems in both air and inert atmospheres. The essentially superimposable thermograms point towards a non-oxidative mechanism of degradation for P-T systems. This implies better prospects for application of this type of resin for thermo-structural uses in aerospace in place of conventional phenolics. Laser ablation studies on a series of ablatives including PT resin have confirmed this.

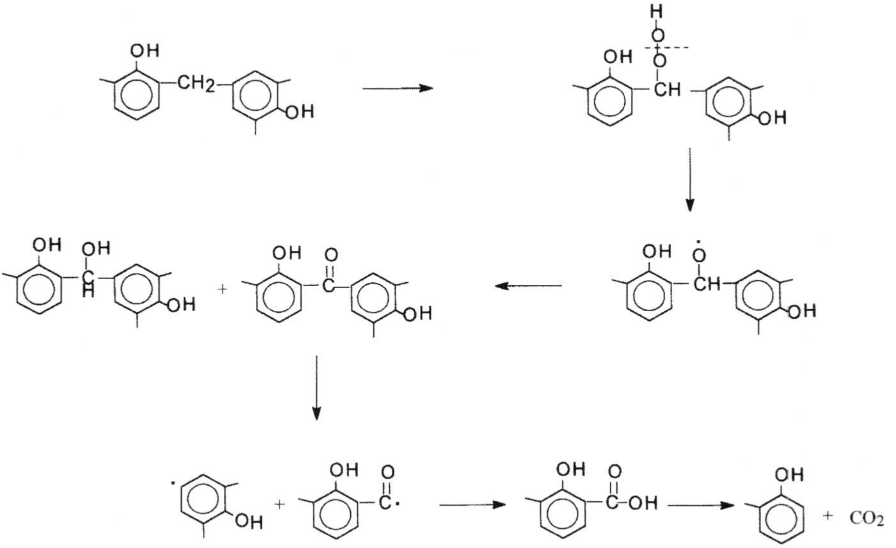

Scheme 7. Oxidative degradation mechanism of phenolic resins

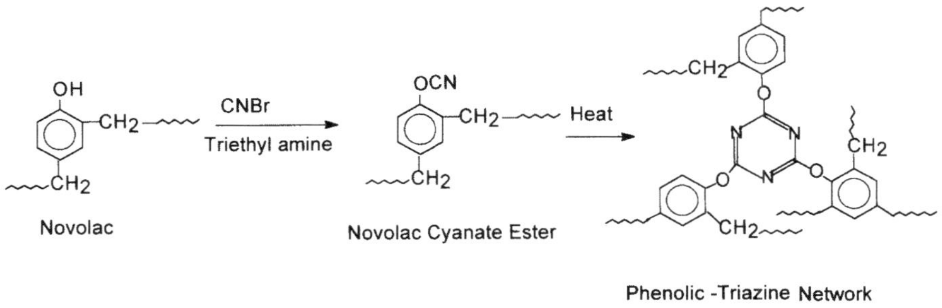

Phenolic -Triazine Network

Scheme 8. Synthesis and curing of cyanate novolac

It was found that the ablation energy was highest for the cyanate polymers on exclusion of phthalocyanines among the polymers tested [76]. Ablative formulations for rocket nozzle applications contain P-T resins as one of the components [77]. The P-T resin systems have been successfully employed in filament winding of cylindrical structures such as pressure bottles which retain 83% of their room temperature properties at 288 °C [68, 78]. Their carbon composites have been experimented on for actuators in turbine engines. A NASA report compared the mechanical properties of composite and char residues of 27 modified phenolic resins including PT resins to those of conventional phenolic resin. Cyanate, epoxy, allyl, (meth)acrylate, and ethynyl derivatives of phenolic oligomers were reviewed. Novolac cyanate along with propargyl-novolac resins provided anaerobic char-yield of 58% at 800 °C [79].

PT resins subjected to sheet molding exhibited onset of decomposition (Ti) at 475 °C, char yield of 62% at 900 °C and good thermal and thermo-oxidative stability [80]. Despite many claims about the superiority of PT systems over other cyanates and phenolics, there are only a few reports on its commercial utilization. One reason, as it appears to us, is that the resin generally shows inconsistency in cure behavior due to catalysis by the spurious impurities adsorbed on the polymer during its synthesis. It is not easy to purify the polymer scrupulously, to the level of monomeric dicyanates. Absorbed moisture can also cause variations in cure behavior. Unpredictable cure profiles imposes processing difficulties and large property variations. It is generally found that the synthesized resins exhibit poor shelf life, particularly when the precursor novolac possesses higher molar mass. P-T resins, structurally modified with both rigid and flexible groups in their backbone were not helpful in improving either shelf life or thermal stability. Thus, a flexible pentadecenyl group (i.e., cardanol) was introduced through copolymerization of phenol with cardanol, and subsequently using the modified novolac for cyanation. The thermal stability decreased proportional to the cardanol content and the resins exhibited poorer shelf-life [48]. The crosslinked cyanate ester of polymaleimidophenol also showed inferior initial decomposition properties, although the char yield was significantly higher [81, 82]. Poly maleimidophenol was formed by the thermal polymerization of *N*-(4-

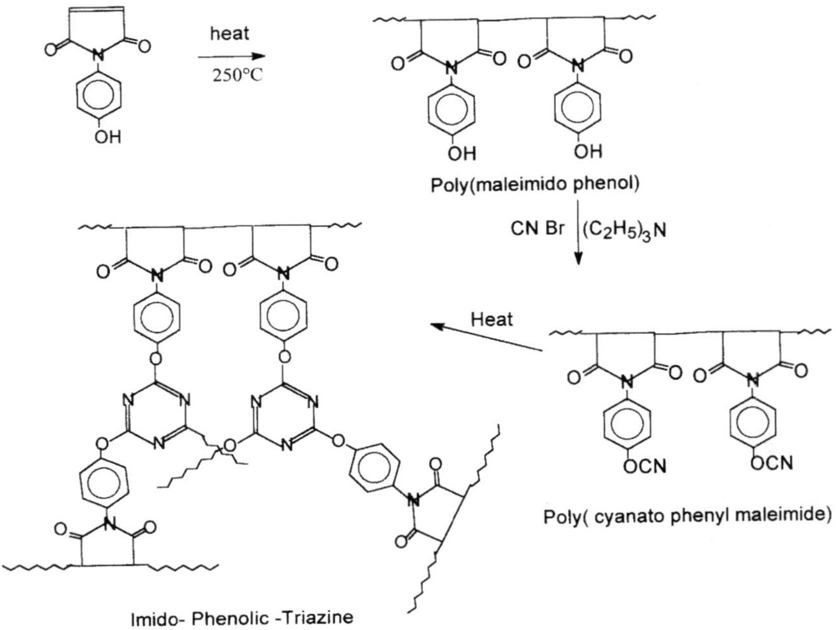

Scheme 9. Synthesis of imido-phenolic-triazine net work polymers

hydroxy phenyl) maleimide (HPM) and was transformed to the cyanate ester. Thermal curing of the latter gave the imido phenolic triazine as depicted in Scheme 9.

The cyanate ester of an imide-modified novolac was also synthesized [83]. The imide-modified novolac was derived from copolymerization of phenol and HPM with formaldehyde in the presence of an acid catalyst [84]. The cyanate ester, synthesized from the phenolic resin by the conventional route, underwent a two-stage cure, implying independent cure of both the cyanate and maleimide groups. However, the cured imido phenolic triazine exhibited poorer thermal stability and anaerobic char residue, attributed to the interference of the rigid imide groups in the char-forming reactions of the triazine groups at higher temperature. The structure of the cyanate ester and the imido phenolic triazines are shown in Scheme 10. The scheme has neglected the presence of very minor concentrations of methylol groups present in the parent phenolic resin. A recent patent claims preparation of a low molar mass novolac cyanate ester prepolymer (M_n=310 g/mol) from the corresponding novolac resin. The prepolymers are claimed useful as coatings, adhesives, and as matrix in copper clad laminates for printed circuit boards [85]. Reports on cyanate esters of other polyhydric phenols are few. Different grades of poly (4-cyanato styrene) (PCS, normal, polymer and novolac grades) and copolymers of 4-cyanato styrene with butadiene (PCS-BD) or MMA (PCS-MMA) as comonomers have been prepared by Gilman et al.

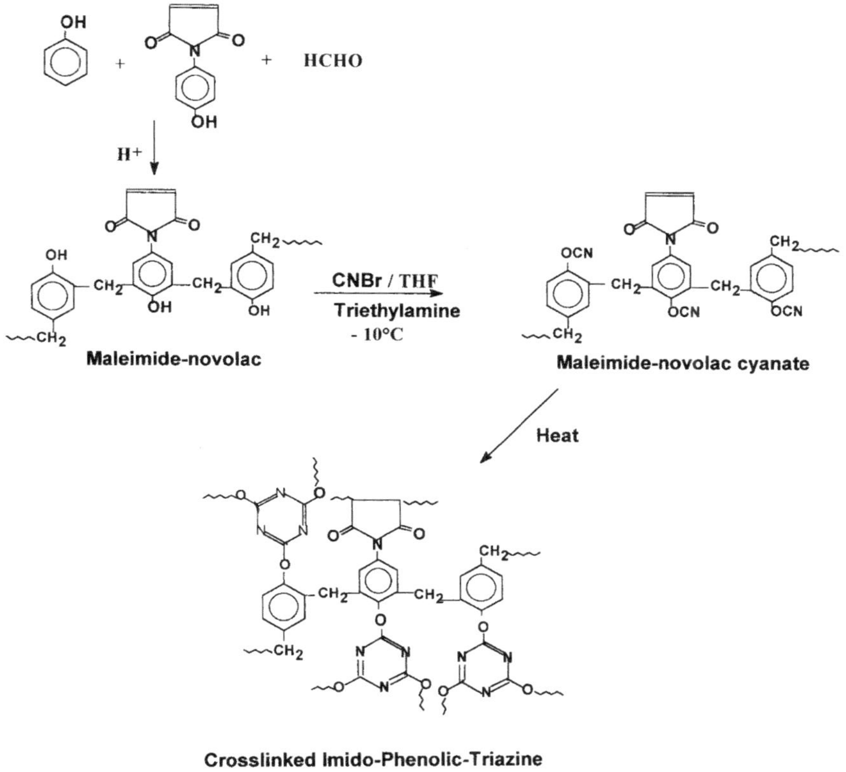

Scheme 10. Synthesis of maleimide-modified cyanate ester and imido-phenolic-triazine

[86]. Flammability tests, performed using a microscale combustion calorimeter, showed significant differences in the flammability of the cured polymers. The flammability decreased with increasing branching of the cyanatophenyl styrene. The best results were obtained for novolac grade polycyanatophenyl styrene. Copolymer PCS-BD showed similar properties to poly cyanato styrene, probably through crosslinking of the unsaturated monomer at high temperature. The PCS-MMA copolymer showed the least flame resistance. On a comparative scale, the PT resins exhibited the best flame resistance. The thermal properties of these polymers, which could have given useful information on their flammability behavior, were not discussed. The structures of the various polymers are shown in Scheme 11. In a related work, copolymers of styrene with 4-vinyl phenyl cyanate or 2,6-dimethyl-4-vinyl phenyl cyanate were prepared via free radical polymerization [49]. The copolymers were sensitive to UV light and crosslinked on irradiation with 254 nm UV radiation. Interestingly, the cyanate groups in the latter copolymer underwent rearrangement to the isocyanate during irradiation, whereas both the copolymers yielded cyanurate networks on thermal curing [49].

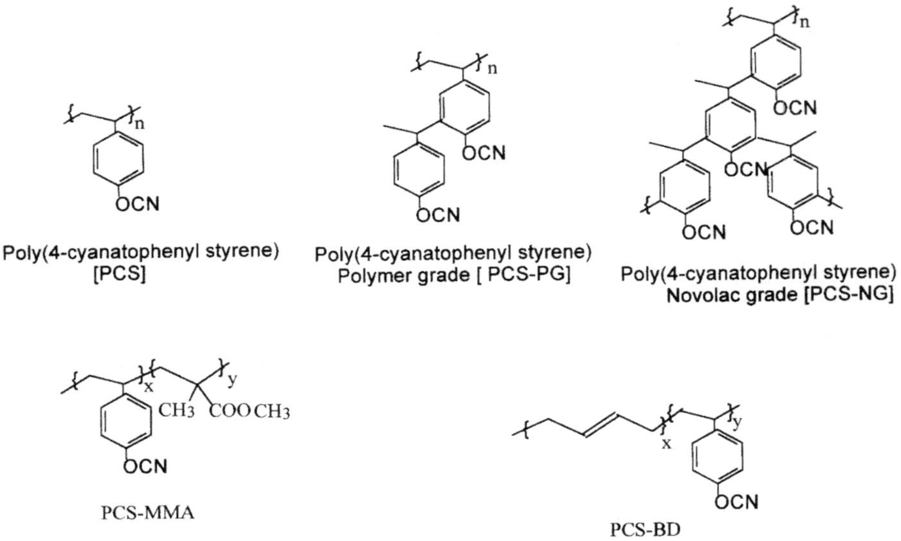

Poly(4-cyanatophenyl styrene)
[PCS]

Poly(4-cyanatophenyl styrene)
Polymer grade [PCS-PG]

Poly(4-cyanatophenyl styrene)
Novolac grade [PCS-NG]

PCS-MMA

PCS-BD

Scheme 11. Structure of different grades of cyanatophenyl styrene copolymers

4
Thermal Curing and Cure Monitoring

Cyanate esters can undergo thermal or catalytic polycyclotrimerization to give polycyanurates with a high degree of efficiency. The catalysts are usually Lewis acids and transition metal complexes or amines. Generally they are cured with a transition metal catalyst or chelate catalyst in the presence of a hydrogen donor such as nonyl phenol. Although the technology based on cyanate ester has taken large strides, the very basic information on its cure mechanism remain disputed. It is generally accepted that the cyanate cure takes place through cyclo trimerization to the polycyanurate. However, evidence for this mechanism is inconclusive to a large extent. The thermal cure reaction of bisphenol A dicyanate (BA-CY) studied by Gupta and Macosko [87] using a mono functional model compound, namely 2-(4-cyanatophenyl), 2-phenyl propane, suggested trimerization to be the major reaction (>80%). Side reactions were also detected, including formation of a dimer and higher oligomers but this occurred only to a minor extent. It appears that the actual mechanism is catalyst-dependent. In the presence of metal ions, the cyanate groups are believed to be coordinated through the metal ion to allow ring closure through a step growth or ionic path [88] (Scheme 12). Normally a co-catalyst like nonyl phenol is also employed. The latter's role is to act as active hydrogen donor. But there is no report concerning the evidence for metal co-ordination with cyanate group. Many researchers have proposed the possibility of formation of cyclic dimer intermediates. Fang and Shimp proposed the diazacyclobutadiene derivative (shown in Scheme 15, later)

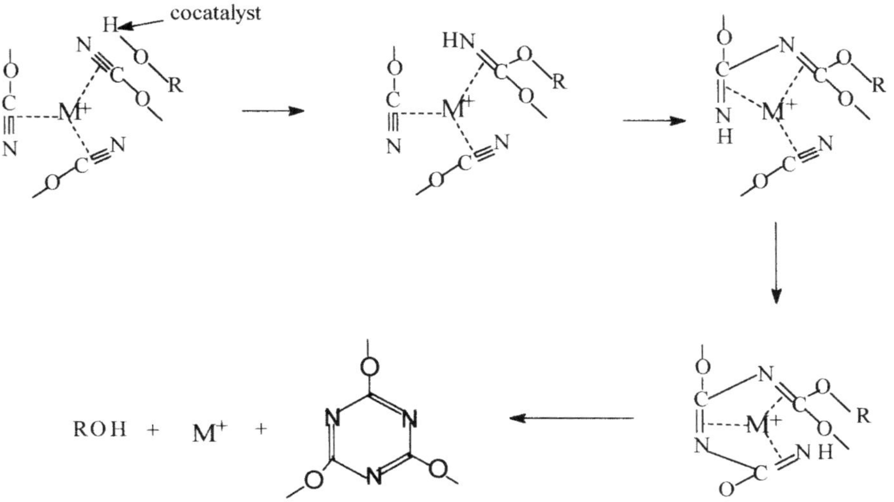

Scheme 12. Proposed mechanism for the metal ion catalysed polymerisation of cyanate ester (Shimp etal)

[89]. The ^{13}C and ^{15}N spectroscopic studies on polymerization of cyanates also could not furnish conclusive proof for the existence of cyclic dimers [90, 91]. Mass spectral evidence has been presented by Fang and Houlihan for the existence of hydrated cyanate dimer, believed to be the actual catalyst for the uncatalyzed polymerization of cyanates [92]. This hydrated dimer mediated mechanism has gained more support with evidence for its presence being presented by researchers. On these lines, Simon and Gillham [93] proposed a reaction and kinetic model (Scheme 13), considering the reaction to be triggered by adventitious water and phenol impurities (whose reaction with the cyanate ester is an equilibrium reaction). Catalysis by the added metal ions, which stabilizes the imino carbonate intermediates by complexing, is also taken into account. The model has considered all possible reaction paths and intermediates.

Grenier-Loustalot et al. [94, 95] investigated the mechanism of polymerization of mono and dicyanates by HPLC and spectroscopic methods. Using HPLC/UV, secondary products like phenolic derivatives of the cyclic trimer and pentamer and carbamates could be detected. The authors proposed formation of dimer, which could inhibit the polymerization and delay gelation. It was found that in catalyzed systems, iminocarbonates, four membered rings (diaza cyclobutadiene derivative) and carbamates were present in non-negligible proportions. In catalyzed systems, the same conversion is obtained at the gel point, and the products are also the same although the reaction is favored. A reaction mechanism as well as a kinetic model close to that of Gillham was proposed (Scheme 14). This model agrees with the findings of Fang and Shimp [89] and also accounts for the observation made by Gupta and Macosko [87], but contradicts the finding of Fyfe et al. [91] who, based on NMR studies, claimed that

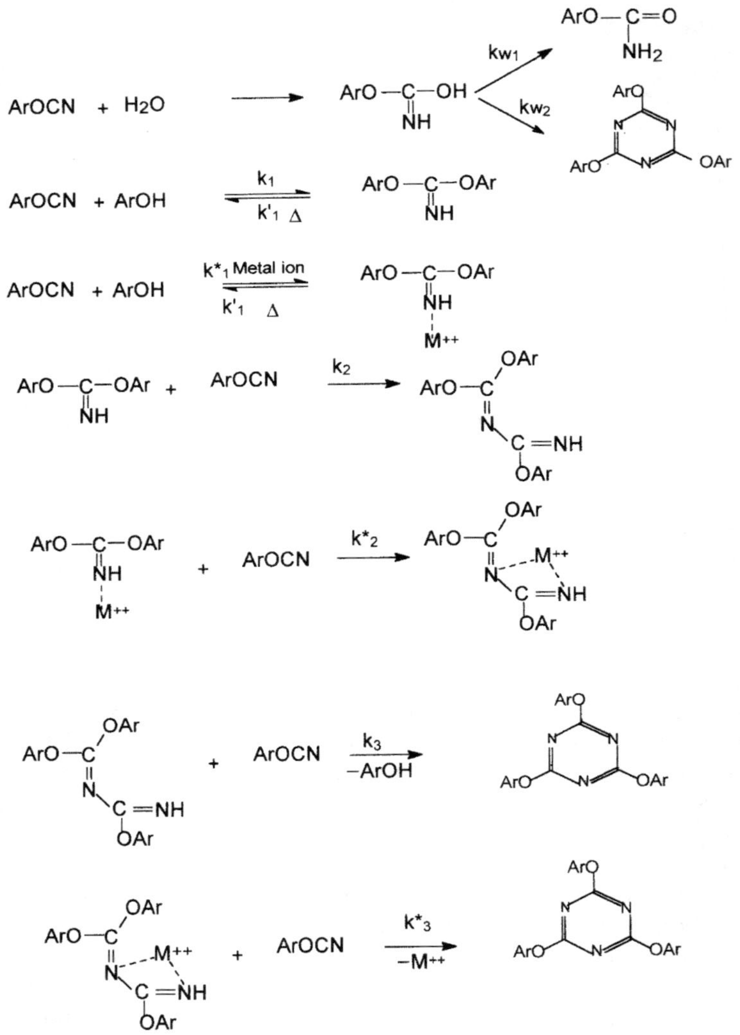

$ArOCN + H_2O \longrightarrow$ $ArO-\underset{NH}{\overset{||}{C}}-OH$ $\xrightarrow{kw_1}$ $ArO-\underset{NH_2}{\overset{|}{C}}=O$

$\xrightarrow{kw_2}$ triazine (ArO, ArO, OAr)

$ArOCN + ArOH \xrightleftharpoons[k'_1 \ \Delta]{k_1}$ $ArO-\underset{NH}{\overset{||}{C}}-OAr$

$ArOCN + ArOH \xrightleftharpoons[k'_1 \ \Delta]{k^*_1 \ \text{Metal ion}}$ $ArO-\underset{\underset{M^{++}}{|}}{\overset{||}{\underset{NH}{C}}}-OAr$

$ArO-\underset{NH}{\overset{||}{C}}-OAr + ArOCN \xrightarrow{k_2}$ product

$ArO-\underset{\underset{M^{++}}{|}}{\overset{||}{\underset{NH}{C}}}-OAr + ArOCN \xrightarrow{k^*_2}$ product

$ArO-\overset{OAr}{\underset{N}{C}}\cdots{C=NH}\,(OAr) + ArOCN \xrightarrow[-ArOH]{k_3}$ triazine

$ArO-\overset{OAr}{\underset{N}{C}}\cdots M^{++}\cdots{C=NH}\,(OAr) + ArOCN \xrightarrow[-M^{++}]{k^*_3}$ triazine

Scheme 13. Simon-Gillham model for polymerisation of dicyanate ester

trimerization is the only reaction product. The same team investigated the molten state reactivity of three different cyanate esters, BACY, BECY, and hexafluoro bisphenol A dicyanate using DSC, NMR, and HPLC and showed trimerization as the major phenomenon, but not the exclusive reaction. A compound with triazine at one end and phenol at the other was identified in one case. The reaction path and products were found to be dependent on the purity level of the monomers. Use of liquid chromatography and UV led to identification of cyclic trimers or pentamers with one to four hydroxyl groups, in a related study [96].

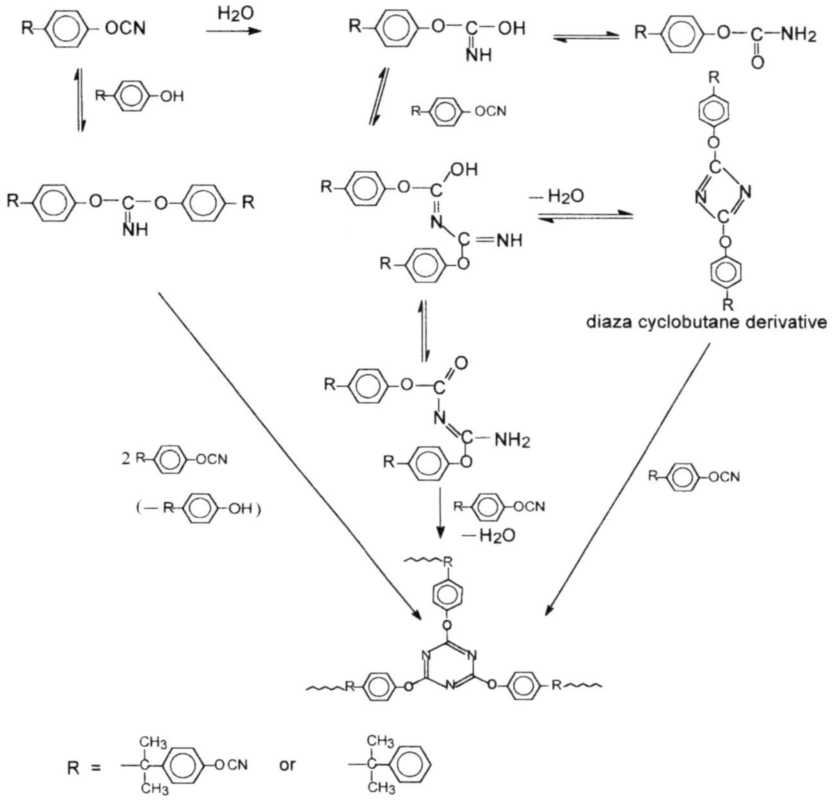

Scheme 14. Reaction mechanism for the polymerisation of cyanate ester proposed by Loustalot et al.

A plausible mechanism for the catalyzed polymerization of cyanates seems to be similar to the one proposed by Brownhill et al. for the $TiCl_4$-catalyzed polymerization [97]. The transition metal co-ordinates with the nitrogen, enhancing the electrophilicity of the nitrile group for addition of another cyanate group on to it as shown in Scheme 15. The network formation and molecular species distribution in the pre-gel region have been studied by a number of techniques like Size Exclusion Chromatography, Differential Scanning Calorimetry [98–100], Nuclear Magnetic Resonance, and dielectric analysis [98], and diffusion controlled kinetic models were derived. The reactions were usually carried out in presence of catalysts like zinc naphthenate, zinc octanoate [101], cobalt naphthenate, and nonyl phenol [102], etc. The models are capable of explaining experimental conversions over the entire range of cure. In addition to the above-mentioned catalysts, $CuCl_2$ (in acetone) which forms a 2:1 complex with bisphenol A dicyanate followed by its cyclotrimerization [103] and $TiCl_4$ (in dichloromethane)

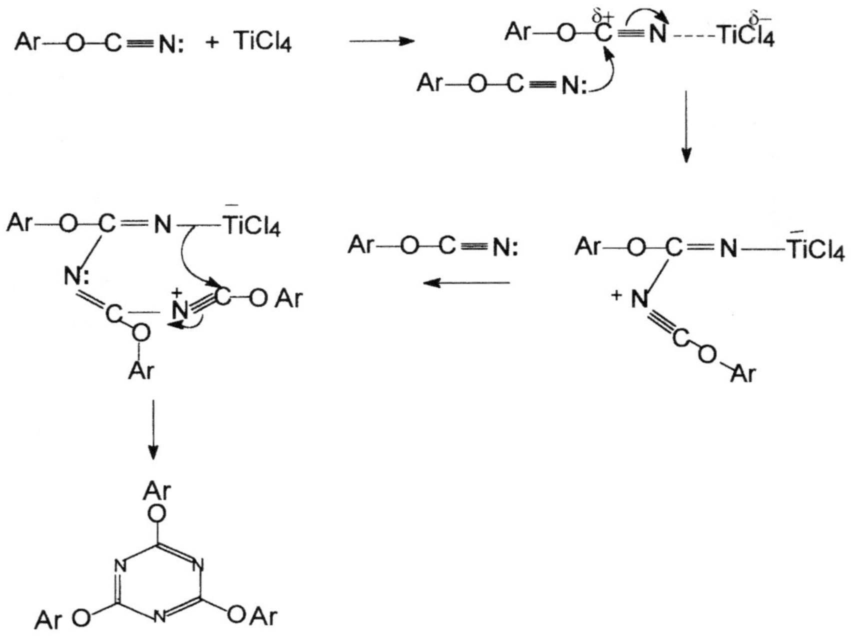

Scheme 15. Brownhill's mechanism for TiCl$_4$-catalysed trimerisation of cyanate ester

capable of forming cyanate-TiCl$_4$ complex [97] are claimed to be efficient catalysts for the cyclotrimerization of cyanate groups at ambient temperature..

Among the advanced techniques employed to follow the cure reaction, Fiber Optic Raman Spectroscopy has been an effective tool. By this technique, both the temperature build-up and the cure advancement of AroCy L-10 could simultaneously be followed. The local temperature of the sample, determined by Raman-Stokes and anti-Stokes scattering of a reference peak correlated well with the temperature measured using a thermocouple probe. The extent of cure could be monitored using either individual peaks associated with the reactant or product or by using the entire spectrum [104].

5
Polymerization Catalysis and Kinetics

To understand and control the processing of this new generation thermoset, it is imperative to study the kinetic aspects of its cure reaction, as they drive the complex changes in morphology and structure of the polymer during its processing operations. There have been a number of reports on the kinetics of the catalyzed network formation of dicyanate esters including BACY. The processing of the resin has also been subjected to extensive studies. The cure reaction is highly susceptible to catalysis by a large variety of materials including transition metal carboxylates, acetyl acetonates, phenols, metal carbonyls, adventitious water, etc. [45, 89, 105–110]. The studies on kinetics of the catalyzed reaction have also been reported [102, 106, 111–16].

Mechanism and kinetics of polymerization of various mono and difunctional CEs have been studied by techniques employing HPLC, liquid and solid state ^{13}C NMR, UV, and FTIR spectroscopy [96, 117–19], dispersive fiber optic Raman Spectroscopy [104], DSC, and SEC [98, 99], rheological and dielectric measurements [98, 109], etc. Torsional braid analysis has been particularly useful for cure advancement. [93]. ^{15}N NMR is a reliable tool for cure monitoring of cyanate esters as the signals due to the OCN and triazine appear as two isolated signals at around 215–226 ppm and 180 ppm, respectively [90]. Galy et al. studied the effect of the catalyst concentration on gelation of cyanate esters using DSC and viscometry [102]. The Co^{2+}-nonylphenol catalyst in different concentrations was used for three different monomers, namely BACY, BECY, and hexafluoro bisphenol A dicyanate (BFCY). For the same proportion of catalyst per cyanate function, the order of reactivity for the cyanate monomers was reversed for the catalyzed and uncatalyzed systems. The least reactive monomer in the uncatalyzed state became the more reactive one on catalysis. Increasing catalyst concentration decreased the ΔH of the reaction, implying a possible change in mechanism on enhancing catalyst concentration. The cured network showed β-relaxations at −100 °C, 70 °C, and 180 °C, wherein the latter two could not be assigned. The decrease in heat of reaction with increased catalyst concentration has been observed by us as well [115] in a study to be discussed later. A similar catalyst effect on differences in cure enthalpy has been reported by Simon [120]. This interesting observation needs further investigation.

As stated earlier, the cure reaction is influenced by a number of factors like impurities and environment [121, 122], solvent [117], catalysts [95], etc. It is reported that no reaction occurs if absolutely pure CE is heated [107]. In the absence of an externally added catalyst, the reaction is believed to be catalyzed by adventitious water and residual hydrogen donating impurities like phenol [122] as stated above. The catalysts for cure generally include imidazole and transition metal complexes [95, 98, 102, 109]. The presence of nonyl phenol co-catalyst has been responsible for lowering T_g of the final cured network due to plasticization effects [109]. Among the adventitious impurities, moisture plays a significant role in kinetics of curing, hydrolysis of products, gelation, etc. [123, 124]. It was found that T_g of the final network could also be lowered due to plasticization when cured in a solvent [117]. Cyanate esters can be polymerized under photochemical conditions indicating their prospects for application in high T_g photoresists. The kinetics of photocatalyzed polymerization in the presence of tricarbonyl cyclopentadienyl manganese was followed by DSC and FTIR techniques. It was found that the network developed higher T_g at lower conversions than that observed in uncatalyzed polymerization. The reaction was first order with respect to concentration of both cyanate and catalyst. The proposed mechanism involves a photosubstitution of the carbonyl group in the catalyst by the cyanate group to form a new complex, which produces an active catalyst on thermal activation. The activation energy decreased with irradiation time [108].

The kinetics of thermal and photochemical polymerization has been monitored by employing phosphorescence probes too. These organometallic photoluminescent probes are highly sensitive to viscosity changes of over five orders of magnitude within the polymer network [125]. An organometallic complex of cobalt was used to study the polymerization of AroCy L-10 and AroCy B-10. The luminescence, arising from a low lying metal to ligand charge transfer, gives a weak emission band at 580 nm which intensifies and shifts to 565 nm on thermal curing of the cyanate groups [126]. Similar shifts in the emission band at 580 nm has been observed with other organometallic probes as well. Thus a rhenium complex has been successfully used to study the polymerization of AroCy L-10 and AroCy B-30 [127].

In a majority of the above cases, kinetics is indirectly studied by the gelation behavior of the resin, based on which, researchers have found that transition metal acetyl acetonates are preferable to carboxylates or naphthenates since the resulting networks have better thermal stability and lesser moisture absorption property in the former case [106]. The catalytic efficiency of manganese acetyl acetonate has been found to be the maximum. Gelation has been the parameter monitored for comparing the catalysis by a few transition metal acetyl acetonates as studied by Gupta and Macosko [113] and organic and organometallic compounds as studied by Korshak et al. [114].

Transition metal organometallic complexes like dicarbonyl cyclopentadienyl iron [128], tricarbonyl cyclopentadienyl manganese [129] and iron-arene complexes [130, 131] have also been reported as photoinitiators for photochemical crosslinking of cyanate esters. Photosubstitution of carbonyl groups by -OCN during irradiation initiates the reaction in the former case whereas photochemical dissociation of arene triggers it in the latter system.

The majority of the studies on catalysis centers around transition metal complexes, particularly acetyl acetonates. Most of the reported work on the catalysis of cyanate cure gives unreliable and at times conflicting results because of the imperfection in the methods employed and the uncertainties of the kinetic models assumed [100, 102, 106, 112–16]. With the objective of generating quantitative information on their relative catalytic efficiency in terms of the kinetic activation parameters and to establish the correlations between the catalytic efficiency and the characteristics of transition metal acetyl acetonates, a detailed investigation of the kinetics of thermal cure of BACY by these materials has been performed by us [115]. The method is based on dynamic DSC. Here a new kinetic approach was made for the thermal cure reaction of bisphenol A dicyanate in presence of various transition metal acetyl acetonates and also dibutyl tin dilaurate [DBTDL]. The cure reaction was found to involve a pre-gel stage corresponding to around 60% conversion and a post-gel stage beyond that. The reaction could be satisfactorily assigned an overall order of 1.5 in contrast to previous studies assigning second order auto catalytic model. The activation energy (E), pre-exponential factor (A), and order of reaction (n) were computed by the

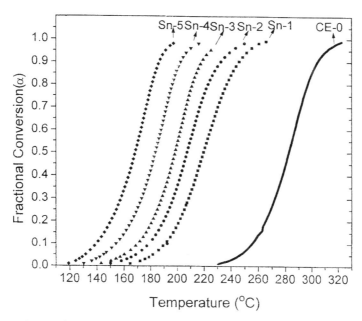

Fig. 4. Dependency of time-conversion profile for the cure of BACY at varying concentrations (mmol/mol of cyanate) of DBTDL from DSC. Sn-1: 0.055, Sn-2: 0.11, Sn-3: 0.22, Sn-4: 0.44, Sn-5: 0.88, CE-0: no catalyst, heating rate :10 °C/min

Coats-Redfern method [131a]. A kinetic compensation (KC) correction was applied to the data in both stages to normalize the E values. The normalized activation energy showed a systematic decrease with increase in catalyst concentration. The cure temperature shifted to a lower temperature on increasing the catalyst concentration as demonstrated in Fig. 4 for the case of catalysis by ferric acetyl acetonate. An exponential relationship was observed between the KC-corrected E and catalyst concentration, substantiating the high propensity of the system for catalysis. This relationship is shown in Fig. 5 for the two cases of ferric acetyl acetonate and DBTDL. Since the cure was characterized a pre-gel and post gel sector with difference in kinetic behavior, the analysis was done separately for the two sectors as shown in this figure. At fixed concentration of the catalyst, the catalytic efficiency, measured in terms of the decrease in E value, showed dependency on the nature of the coordinated metal and stability of the acetyl acetonate complex. Among the acetyl acetonates, for a given oxidation state of the metal ions, E decreased with decrease in the stability of the complex. Thus, the catalytic activity increased in the order $Mn^{+2}>Ni^{+2}>Zn^{+2}>Cu^{+2}$ for the divalent ions and in the order $Fe^{+3}>Co^{+3}>Cr^{+3}$ for the trivalent metals, which in both cases is the reverse of the order of stability of the complexes. Manganese and iron acetyl acetonates were identified as the most efficient catalysts in league with the observation of others using different techniques. A linear re-

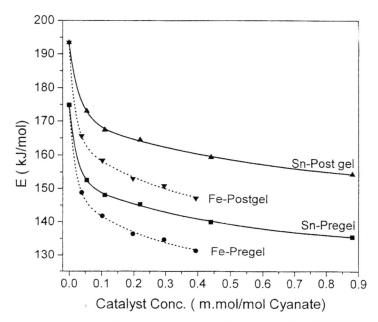

Fig. 5. Variation of normalized activation energy with concentration of DBTDL and ferric acetyl acetonate for the catalyzed polymerization of BACY

lationship was found to exist between the KC-corrected activation energy and the gel temperature for all the systems. The latter was inversely proportional to the catalyst concentration for a given catalyst. This relationship is shown in Fig. 6 for various transition metal complex catalysts for the two kinetic sectors. During the course of investigations on the cure of acrylate monomer/cyanate ester blend by DSC, Hamerton examined the effect of a few metal ions in polymerization of AroCy L-10 and AroCy XU366 [132]. The metal ions, when employed at same molar concentrations, ranked their catalytic efficiency in the order, $Zn^{2+}>Cu^{2+}>Co^{2+}>Ti^{2+}>Al^{3+}$ measured in terms of the onset temperature of AroCy L-10. The order changed to $Zn^{2+}>Ti^{2+}>Cu^{2+}>Co^{2+}>Al^{3+}$ when the monomer was changed to AroCy XU366. When the catalyst concentration is a few hundred ppm, the enthalpy of polymerizations was more or less the same for all the catalyzed systems (90 kJ/mole of cyanate) which was slightly lower than that for the uncatalyzed monomer (97–102 kJ/mol cyanate) [133].

Earlier studies attributed first-order kinetics to the reaction, based on evidence from on-line IR and DSC analysis [116], the reaction presumably being induced by phenolic impurities. The deviation from mean field theory is attributed to diffusional limitations, which is more prominent in CEs with rigid backbone.

A better empirical kinetic model consisting of two parallel, second order competing reactions, one of which is autocatalytic towards the dicyanate, was proposed by Simon and Gillham [93] for the uncatalyzed reaction.

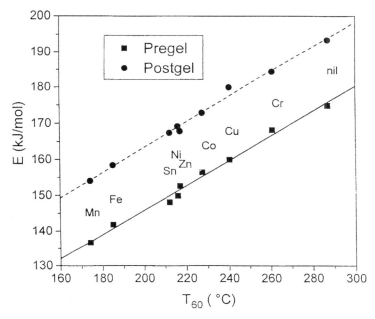

Fig. 6. Linear relationship of activation energy (E) with gel time (T_{60}) for different catalyst systems for polymerization of BACY

$$dx/dt=k_1(1-x)^2+k_2x(1-x)^2 \tag{1}$$

where x is the conversion.

The model satisfactorily described the cure behavior for the entire range as experimentally monitored by FTIR, DSC, and torsional braid analysis (TBA). This model satisfactorily explained the cure behavior of both catalyzed and un-catalyzed systems over a wide range of temperature and throughout the curing process. The authors proposing the kinetic model considered the reaction to be triggered by the adventitious water and phenol impurities (whose reactions with the cyanate ester is considered as an equilibrium reaction). Catalysis by the added metal ions, which stabilizes the imino carbonate intermediates by complexing, is also considered. The model has considered all possible reaction paths and intermediates as detailed in Sect. 4 and depicted in Scheme 14. Considering the various reactions, expressions could be obtained for the individual apparent empirical rate constants of the second order autocatalytic model in terms of the actual rate constants and equilibrium constants.

Other researchers have also successfully applied the autocatalytic model [122, 134]. For fitting the conversion-time curves in this context, the conversion can be indirectly obtained from the T_g of the partly cured matrix. Georjon et al. [122] studied the evolution of T_g with cyanate conversion in the isothermal cure of uncatalyzed BACY. A one-to-one correlation existed between conversion and T_g, correctly described by a restated DiBeneditto's equation.

The T_g of a partially cured network is originally given by the DiBenedetto's equation [135] as

$$T_g = T_{go} + (T_{g\alpha} - T_{go})\lambda x / [1 - (1-\lambda)x] \tag{2}$$

where T_{go} and $T_{g\alpha}$ are the glass transition temperatures of the monomer and the fully cured network respectively. λ is an adjustable parameter. Pascault and Williams [136] restated the DiBenedetto equation taking $\lambda = Cp_\alpha/Cp_o$, where Cp_o and Cp_α are the heat capacity changes during the glass transition of the monomer and the network, respectively. Since the network cannot undergo the maximum theoretical extent of conversion [i.e., 100%] due to diffusional limitations, the experimental data deviated from the above equation beyond the gel point. In this case, a reasonably good correlation could be found when x was rationalized as x' and λ as λ', where $x' = x_m/x$ and $\lambda' = Cp_m/Cp_o$, where m refers to point of maximum cure in the practically cured network. This point was considered as corresponding to 95% conversion and the restated T_g-conversion equation [137] is

$$T_g = T_{go} + (T_{gm} - T_{go})\lambda' x' / [1 - (1-\lambda')x'] \tag{3}$$

This equation conformed well to the experimental T_g-conversion profile studied by Georjon et al. [122]. The kinetics of dicyanate cure was satisfactorily modeled with two competing reactions, second order and second order autocatalytic, as originally proposed by Simon and Gillham (Eq. 1 above). Conversion was obtained from the T_g data determined by DSC. A test of the validity of the above kinetic model was accomplished by comparing the experimental and calculated time required for the T_g to attain a specified value at a specified temperature using an integrated form of Eq. (1). The agreement between experimental and predicted T_g-time profile was seen only till the point of vitrification. The possible change of the kinetics from the chemically controlled one to the diffusion controlled one in the vicinity of the isothermal vitrification accounted for the observed deviation. During this investigation, it was also observed that T_g decreases for longer cure at higher temperature (250 °C) due to thermal degradation as has been previously reported [138]. It was impossible to reach 100% conversion because of diffusion limitations and the concomitant degradation associated with high temperature cure.

The slightly altered form of the Pascault-Williams equation [136] was employed by Venditti and Gillham for describing the conversion-T_g dependency for many thermosetting resin including cyanate ester [139]:

$$\ln (T_g) = [(1-x)\ln (T_{go}) + \lambda x \ln (T_{g\alpha})] / [(1-x) + \lambda x] \tag{4}$$

where, $\lambda = \Delta_{Cp\alpha} / \Delta_{Cpo}$
$\Delta_{Cp\alpha} = Cp_\alpha^l - Cp_\alpha^g$, corresponding to x = 1 and
$\Delta_{Cpo} = Cp_o^l - Cp_o^g$, corresponding to x = 0

$\Delta_{Cp\alpha} / \Delta_{Cpo}$ can be determined by DSC

Cure kinetics of the liquid dicyanate monomer, AroCy L-10, was followed by DSC and IR studies. Experimental data showed that diffusion control occurs well before vitrification of the cyanate resin. A two-step kinetic model and a WLF type diffusion limited kinetic model were proposed to describe the entire range of the reaction. The WLF-type kinetic model gave better fit for a second order reaction [140]. The influence of substituents on mono- and difunctional cyanates has been studied by Bauer et al. [141]. Thus, a model was developed based on reaction kinetics and network statistics to explain the gelation behavior of monocyanates (gel point around 50% conversion). Pairs of different monofunctional monomers, when trimerized to full conversion, showed distribution of homo and mixed trimers in the final product, indicating the influence of substituents on reactivity. The distribution follows a simple binomial for cyanates with equal reactivity whereas it shows deviation in the case of unequal reactivity. The reactivity of the cyanate ester monomers could be compared qualitatively by a method based on association of the cyanate groups with phenol. The corresponding shift in FTIR absorption of the -OH band of the phenol is correlated to the data obtained from gelation, trimer distribution, and DSC studies.

However, the cure behavior of the cyanate systems has been reported to be different in prepregs, the difference attributed to the presence of fibers. The behavior was studied using dynamic DSC [142], solid state ^{13}C-NMR, SEM, and FTIR [99] for a ^{13}C-labeled BACY/carbon fiber prepreg. At high cure temperatures, the observed T_g decreased for the prepreg.

5.1
Diffusion Controlled Kinetics

In the polycyclotrimerization reaction, the mean field theory usually fails to describe the structure build-up. According to the mean field theory, the degree of polymerization can be defined in terms of the cyanate conversion by the relation

$$DPw=[1+2p]/[1-2p] \tag{5}$$

where p is the conversion.

According to this, gelation is expected to occur at 50% conversion. However, all experimental evidence points to the fact that diffusion limited, reduced reactivity of the OCN functions during later stages of cure delays the gel point to about 60% conversion. This is particularly true for cyanate esters with rigid backbones. In a system where the cyanate groups were separated by a flexible spacer based on phenylene sebacate, the gel point reached the theoretical value of 50% [143]. In this case, the reaction followed first-order kinetics with an overall activation energy of 21.8 kcal mol⁻. The study, based on the prime method, also showed that activation energy remained practically constant for the 30–70% conversion regime. The values decreased to either side of this conversion regime. Further, in this case, the absence of autocatalysis was attributed to the internal catalysis by the ester groups. In another treatment, a WLF-type diffusion controlled kinetic model fitted well to the experimental conversion profile for the isothermal cure of BACY [140]. Experimental data showed that the diffusion limitations come into play well before the gelation. A kinetic model, which may be considered as refinement of the Simon-Gillham model was applied, as given in Eq. (6):

$$1/k(x,T)=1/k_t (T)+1/k_D(x,T) \tag{6}$$

where k is the experimental rate, k_t the true rate constant, and k_D the diffusion rate constant. The temperature dependence of the k_D could be expressed by the refined WLF equation as

$$\log (k_D/k_{Do})=C_1 (T_c-T_g-D)/[C_2+(T_c-T_g-D)] \tag{7}$$

T_c being the cure temperature and C_1, C_2, and D adjustable parameters. T_g was correlated to conversion in this study, using the Havlicek and Dusek equation [143a], in contrast to the DiBenedetto equation as restated by Pascault et al.:

$$1/T_g=(1-x)/T_{go}+x /T_g+C_\alpha(1-x) \tag{8}$$

Deng and Martin [99] examined the diffusion effect on reactivities and evolved a model for gel point and molecular weight. Modeling on a Zn-catalyzed cyanate system showed that unequal reactivities have a significant effect on gel point which is delayed to a higher conversion. This model is an exact solution for the partitioned cure system and also predicts the evolution of molar mass. Since the predicted gel conversion decreases with temperature while the experimental results are invariant, the authors concluded that localized reactions and cyclizations are also responsible for the gel delay. It is generally accepted that the delayed gelation is due to the unequal reactivities of the cyanate groups, dictated also by the diffusion limitations. However, Lin et al. [143b] suggested that the delayed gel point at the actual conversion of 58–65% rather than at the theoretical conversion of 50% could be due to easy intramolecular cyclization during polycyclotrimerization of dicyanate esters, as normally encountered during the reaction of high functional monomers. The extent of delayed gel depends on monomer structure and accessibility of the functional groups, substitution effect, and extent of intra molecular cyclization. These effects are interdependent. The authors developed theoretical expressions for gel conversion, sol fractions, etc. by use of recursive method with due consideration of the intramolecular cyclization.

A modified kinetic model based on the Rabinowitch [144] approach, taking into account the diffusion phenomena including the molecular diffusion processes and molecular size distribution, has been found to describe the conversion profile of Zn-catalyzed dicyanate cure for the entire range [98]. The average diffusivity decreased by several orders of magnitude during cure. The Rabinowitch model explains the diffusional limitations in reactions of small molecules as

$$1/k=1/k_c+1/k_{d0}D \tag{9}$$

where k is the apparent rate constant, k_{d0} another constant, and k_c the intrinsic rate constant. D is the diffusion coefficient. Substituting the diffusion terms into the rate expression for simple second-order kinetics for cyanate cure, the following second-order diffusion model, not considering the auto catalysis, was obtained by Deng and Martin [98]:

$$dx/dt = \{(k\, k'_{do} D/D_{x=0.25})(1-x)^2\}/\{k+(k'_{do} D/D_{x=0.25})\} \tag{10}$$

where $k'_{d0} = k_{d0} D_{x=0.25}$ is a constant. $D_{x=0.25}$ is the diffusion coefficient at a conversion of 25%.

The average diffusivity was estimated using the dielectric analysis-based approach.

The prediction by this model conformed satisfactorily to the experimental time-conversion profile for the entire range, estimated by FTIR technique.

Different reasons have been attributed to the delayed gel point, but few proposals consider reactions other than trimerization in explaining the gelation. According to Gupta's computations [145], dimerization reaction can contribute to delayed gelation. Thus, if α_1 is the extent of dimerization and α_2 that of trimerization, then

$$DPw = (1+\alpha_1+2\alpha_2)/(1-\alpha_1-2\alpha_2) \tag{11}$$

This approach could partially explain the delay in gelation beyond 50% conversion, depending on the extent of dimerization. This hypothesis is of relevance, since later studies have indicated the possibility for dimerization during cyanate cure.

6
Gelation and Vitrification

Gel time and gel temperature are two important parameters for the processing of thermoset resins. Because cyanato functional polymers are self-crosslinking, their gelation studies especially interest researchers. Gelation occurs at the point when the molecular motion ceases on a macroscopic level and it determines the point for application of pressure for compaction while molding the resin, particularly in composites. Both the temperature and catalyst decide the gel phenomenon. Figure 7 shows the dependency of gel time on temperature and catalyst concentration for the thermal cure of BACY in the presence of DBTDL [146]. The gel time is related to temperature by an Arrhenius-type relationship. The effect of various solvents on the gelation of BACY for its solution polymerization has been reported, and the conversion was highest when highly polar solvents like nitrobenzene were used [114]. Various catalysts affected the gel conversion to different extents in good solvents. At low monomer concentrations, the conversion at gel point decreased by 15% and the dependency of gel point on monomer concentration was linear in these ranges. Gelation can be easily studied by DMA technique by monitoring the viscoelastic behavior. Figure 8 shows a typical DMA behavior of BACY using single ply glass prepreg. The gelation is observed by the sharp increase in storage modulus (E') and the glass transition of the cured matrix accounts for the fall E' at higher temperature [146]. An extension of the dynamic mechanical analysis from measurement of melt elasticity was

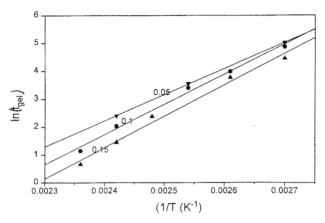

Fig. 7. Dependency of gel time on catalyst concentration for the DBTDL-catalyzed polymerization of BACY

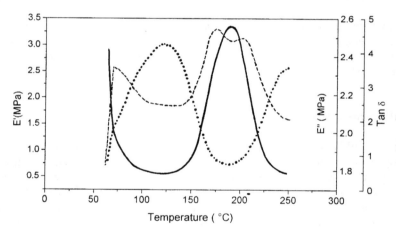

Fig. 8. DMA (non-isothermal) of BACY-glass fabric (prepreg). (——): E′; (- - - -): E″; (●●●●●●): tan δ. Heating rate 2 °C/min, 1 Hz

used to study the novolac cyanate resin vitrification process [147]. The degree of cure obtained was correlated with that from DSC. The disks, prepared by compression molding of plaques of polyethylene (75%) and the PT resin (25%) at ~150 °C, were analyzed for melt elasticity at an isothermal cure temperature of 250 °C. The method proved reliable.

7
Processing

Like other thermoset resins, cyanate esters are amenable to processing by a large variety of conventional techniques and their processing versatility in contrast to the BMI and PMR resins gained them widespread acceptability in composites for a variety of applications. The flexibility is further enhanced by blending with other resins such as epoxies, BMIs, additives, toughening agents, etc.

Cyanate esters are processed by heating, without any need for pressure during molding. A catalyst is normally recommended, preferably a transition metal complex in the presence of nonyl phenol as the co-catalyst. The alkyl phenol not only helps disperse the catalyst but provides active hydrogen source and the cure rate is also dependent on the latter. The cure cycle depends on the catalyst level. However, it is generally accepted that catalysts are useful for inducing early gelation only. High conversion obligatorily warrants heating to high temperature irrespective of the catalyst level. Like many other thermosets, a rule of thumb for high temperature cure for high conversion is applicable to cyanate esters as well. This is mostly related to the diffusional limitations of the high-conversion cure reactions, as discussed previously. The temperature dependency on conversion is demonstrated in Fig. 9 for the cure of BACY, catalyzed by zinc naphthenate/nonyl phenol system [148]. A direct relationship between cure temperature, conversion, and T_g existed.

Since most of the cyanate esters are crystalline materials, in the majority of the cases the resins are B-staged so that they develop good tackiness. Since cyanate polymerization is accompanied by high heat of reaction, use of B-staged resin helps reduce the risk of highly exothermic, autocatalytic, and violent polymerization. Prepolymerization also reduces the risk of meltflow during processing at higher processing temperatures. BACY is a fine crystalline powder and can be resinified by partial polymerization. Depending upon the extent of

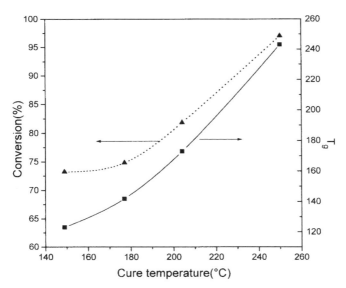

Fig. 9. Dependency of T_g and conversion on cure temperature for BACY, catalyzed by Zn naphthenate (0.15 phr)/nonyl phenol (2 phr) system

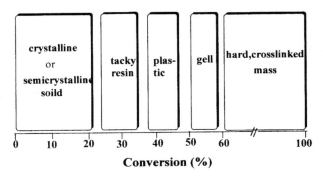

Fig. 10. Physical states of BACY polymerized to various extents

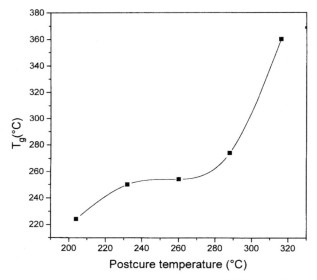

Fig. 11. Effect of post cure temperature on T_g of uncatalyzed PT resin. Cure duration 4 h

cyanate prepolymerization, the consistency of the B-staged resin could vary as shown in Fig. 10. At about 30% conversion the resin attains good tackiness, ideal for fabrication of prepregs. The resinification, uncatalyzed or catalyzed by mild catalysts, can be done in the solid state but control is difficult. It can be conveniently done in solution, which can be directly used for impregnation.

The PT resin already supplied as resinous product can be processed without need for B-staging. A cure temperature of 300 °C is reportedly necessary to achieve high conversion. They generally exhibit T_g higher than 300 °C even at moderate conversion. However, to achieve high T_g, relatively long cure cycles and high cure temperatures close to the DSC exotherm are recommended [148]. Post cure at high temperature has a tremendous influence on the T_g of the cured PT resin as demonstrated in Fig. 11. High conversion is imperative for cyanate esters to achieve not only good mechanical performance but also to impart desirable dielectric properties and solvent and chemical resistance [149]. When it concerns blending of CEs with thermoplastics and thermosets, different options are available. While epoxy, BMI etc can be blended in hot melt form, thermoplastic toughening agents may be mixed in solution.

Thin films of cyanate esters have been photocured using iron arene complexes of the type [CpFe Tol]X$^-$ where X=PF$_6$ or SbF$_6$ [130]. Itoya et al. suggested the possibility of making hard aromatic polycyanurates by applying high pressure during thermal molding of dicyanate monomers [150]. Cyanate ester, AroCy L-10/graphite composites have been successfully pultruded [151]. The mechanical and fracture properties of CEs have been shown to depend on the extent of cure and the resultant network structure. The decrease of Young's modulus with increase in conversion was attributed to reduced cohesive energy density and higher magnitude of sub-T_g relaxations for the more crosslinked system. The unexpected variations in yield stress and strain were attributed to the fact that the main contribution to macroscopic deformation was inelasticity [152].

7.1
Processing of Composites

In a wide range of applications, cyanate esters are used in the form of composites. The conventional composite processing techniques, including prepregging [146, 153], resin transfer molding [154], filament winding [155], and sheet molding [80], are applicable to cyanate systems as well. The potential to tailor pot life and reactivity to specific processing method also exists for cyanate esters and their blends. The resin can be prepregged onto fibers such as carbon, glass, aramid, quartz, polyethylene, etc. [156]. Moisture absorbed on the fiber can alter the cyanate cure and interphase properties and lead to blistering, although some studies point to the fact that moisture has little effect on ILSS of cyanate composites in contrast to epoxy and BMI [157]. For applications such as for radomes, prepregs and composites are made from polyethylene where the ultimate cure temperature has to be limited. For RTM mode of processing the resin viscosity has to be low. AroCy L-10 and cyanate-epoxy blends are easily processable in this way. Copper catalysts are advised when low temperature and shorter cure times are warranted. CE has been filament wound to form cylindrical structures [155].Cylinders and rings have been filament wound from PT resin using T-300 or T-650 carbon fibers [158]. The mechanical properties of such PT components were either comparable or superior to the PMR components. The SBSS of carbon fiber composites ranging from 53 MPa to 99 MPa (depending on fiber sizing and conditioning) were retained to a remarkable >50% value at 316 °C. Flexural strength of 339 MPa obtained for T-300 rings was kept to almost 90% at 288 °C while the flexural modulus was unchanged. Filament windability of BACY was demonstrated by fabrication of NOL rings by solution or hot melt impregnation of the resin on E glass. The NOL tensile strength and short beam shear stress were comparable to the diepoxide-diamine systems [159]. Another type of

molding compound used commonly, especially in automotive applications, is
sheet molding compound (SMC). Typically, SMC contains resin, filler, fiber, etc.,
sandwiched between two sheets of carrier films. PT resins have been successfully
sheet molded [80]. Such components find application above 260 °C. Foams suitable for composite sandwich construction have been made from CEs [160]. A
foamable film precursor is made by blending CE and additives including foam-
ing agent and controlling the time, temperature, and shear rate parameters and
casting in to a film [161].

BACY confers very good mechanical properties to its composites. The resin
with about 30% cyanate conversion is straightaway used for solution impregnation. With E glass laminate, ILSS values of the order of 70 MPa can be obtained.
The values increase further with UD composites and also with carbon reinforcement [146]. Processing of composites and their properties have been compiled
by Mackenzie and Malhotra [162]. Hot/melt prepregging has been employed for
cyanate composites, where control of moisture and selection of catalysts are critical for desirable composite properties [163, 164]. Cyanate ester and PT resins
can be processed by RTM, and the rapid cure is achieved for the latter through
application of microwave or high energy radiation [165, 166]. Microwave
processing has gained some interest [167].

Modified cyanate ester/carbon fiber composites have been fabricated by a
bleed resin transfer molding technique (BRTM]) which encompasses the features of both resin transfer molding (RTM) and resin film infusion techniques.
Thus, blends of AroCy B-30 resin with different quantities of epoxy functional
butadiene acrylonitrile rubber (which otherwise possessed inappropriate rheological characteristics for conventional RTM) have been processed by this technique. Contradictory to most reported elastomer-toughened systems, presence
of the elastomer did not significantly reduce the T_g and conversion, whereas the
mode I and mode II interlaminar fracture toughness increased with increasing
elastomer content. Improvement in mode I fracture toughness by about 140%
has been claimed for composite fabricated by the BRTM technique [168]. Matrix
modification with different linear polymeric additives bearing pendant phenol,
cyanate, and epoxy functions was attempted in the glass laminate composite of
BACY [146]. The additives are (1) a pendant epoxy functional butyl acrylate-
MMA-acrylonitrile copolymer (EPOBAN), (2) a pendant phenol functional styrene-phenyl maleimide copolymer (SPM), (3) a pendant phenol functional,
butylacrylate-acrylonitrile-phenyl maleimide (BNM), and (4) a pendant cyanate
functional styrene-phenyl maleimide copolymer (BNMC). The structure and
composition of the additive copolymers are given in Scheme 16. EPOBAN being
a proprietary polymer, its composition is not given. The mechanical properties
and fracture energy for delamination of the glass-laminate composites were estimated as functions of the nature and concentration of these additives. Except
for the epoxy functional acrylic polymer (EPOBAN), all other systems adversely
affected the fracture energy for delamination of the composites, mainly due to
matrix plasticization or its embrittlement. With the exception of the styrene-hy-
droxy phenyl maleimide (SPM) copolymer, the other modifiers impaired the

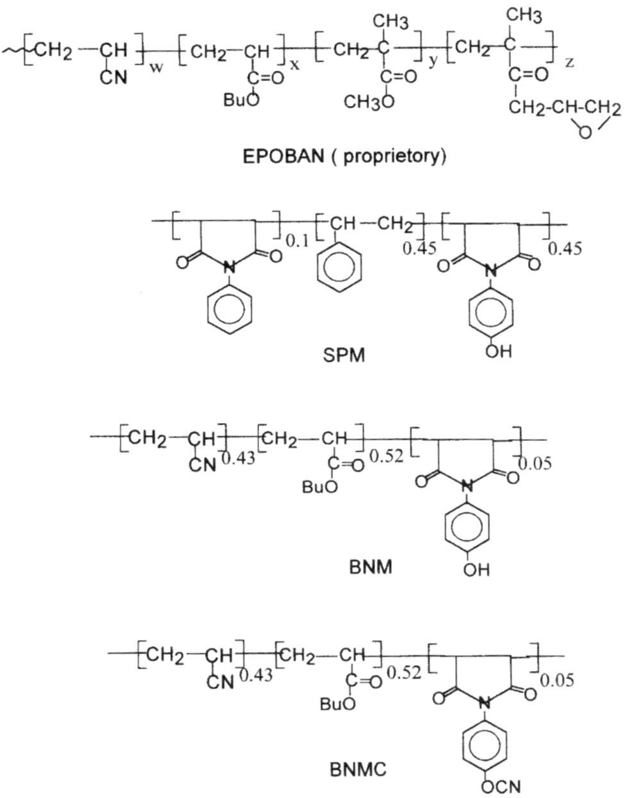

EPOBAN (proprietory)

SPM

BNM

BNMC

Scheme 16. Structure of polymeric additive for modification of BACY matrix

mechanical properties and adversely affected the thermo mechanical profile of the composites. In the cases of a phenol functional acrylic polymer (BNM) and its cyanate derivative, matrix plasticization by the partly phase-separated additive, easing the fiber debonding was found responsible for the impairment of the mechanical properties. The phenol-functional SPM copolymer enhanced the resin/reinforcement interaction, possibly through dipolar interaction induced by the hydroxyl groups. Although this resulted in amelioration of mechanical properties, it led to poor damage tolerance due to resin embrittlement. The mechanical properties of the glass laminate composites are given in Table 3.

DMA analysis substantiated the possible morphological features that could partially account for the trend in mechanical and fracture properties of the glass laminate composites [146]. The DMA, done at a frequency of 1 Hz, are given in Fig. 12, where the percentage retention of the room temperature storage modulus (E') is plotted as a function of temperature to normalize the anomalies resulting from differences in resin content for different systems. EPOBAN-modified

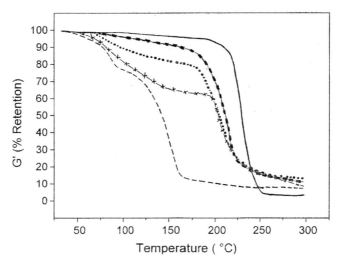

Fig. 12. Dynamic mechanical analysis of BACY-based glass laminate composites. (——): BA-
CY; (-■-■-■-): SPM; (●●●●●●●): BNMC; (-×-×-×-): BNM; (- - - - -): EPOBAN in nitrogen.
Heating rate 5 °C/min, 1 Hz

Table 3. Mechanical properties of glass laminate composite of BACY [146]

Additive	Concentration (wt% of BACY)	Mechanical properties					
				ILSS (MPa)			
		Flexural strength (MPa)	0° Compr. Strength (MPa)	Ambient.	150 °C	Retention at 150 °C (%)	Fracture Energy (kJ/m^{-2})
nil	0	544	295	58	38	66	2.60
EPOBAN	5	401	204	33	24	72	–
	10	503	212	54	31	57	–
	15	501	219	54	17	32	2.50
	20	507	166	40	15	37	2.64
BNM	5	426	240	46	31	78	–
	10	610	220	59	23	39	–
	15	525	260	54	23	43	1.66
	20	420	227	54	24	44	–
BNMC	10	521	257	56	25	45	–
	15	533	262	49	15	31	1.56
SPM	5	542	233	71	40	56	–
	15	598	240	71	50	70	1.31

matrix exhibited two glass transitions at ~80 °C and ~150 °C, corresponding respectively to the acrylate backbone of EPOBAN and the BACY network. A plateau region was observed in between the two transitions. DSC of the composite also showed two glass transitions in this range, confirming that EPOBAN-modified matrix exhibits a heterogeneous morphology in the laminate. EPOBAN is a low-T_g polymer with glass transition below room temperature (~10–15 °C) and it has been observed that polycyanurate has a T_g ~250 °C. However, the blend showed two T_gs in between these two extremes, implying that the two phases are not homogeneous by themselves. The first transition is caused by the elastomer phase containing part of BACY dissolved in it and consequently with a higher T_g. Similarly, the BACY network contains some amount of EPOBAN dissolved in it, as a result of which the T_g of this phase is substantially lowered. Since no co-reaction is expected, the matrix system can be considered as a semi interpenetrating polymer network. The overall lowering of T_g is responsible for the poor high temperature retention of mechanical properties (ILSS and compressive strength) in this case. The plasticization of the matrix, especially at higher concentrations of the additive, is responsible for easing the resin/reinforcement debonding, thereby diminishing the mechanical properties in general. DMA of BNM- and BNMC-modified matrices also showed the presence of phase-separated morphologies. Storage modulii curves show two transitions corresponding to acrylate (of BNM backbone) and triazine backbones occurring at around 90 °C and 200 °C respectively, as can be seen in the DMA spectra (Fig. 12). The rubber phase is not pure and contains dissolved polycyanurate, with consequent enhancement of T_g from 40 °C to 90 °C. The diminution in T_g of the polycyanurate phase is marginal (by ~30 °C). The biphasic behavior and plasticization in case of BNM is anticipated since the phenol groups are not present in adequate concentration to enter into significant network linking with the cyanurate matrix. The cyanate functional polymer (BNMC) completely co-reacts with the continuous phase of BACY. However, the cyanate functions are located statistically in the chain in small concentrations between segments of high molar mass (5000 g/mol). This makes the elastomer segment between crosslinks form a separate phase. As a result of this, the cyanate functional additive also exhibited a biphasic morphology with two glass transitions, one at ~90 °C and another at ~220 °C, corresponding to the elastomer- and triazine-dominated phases, respectively. The low temperature transition is considerably less prominent when compared to its phenol counterpart and the major transition is that caused by the BACY matrix which does not experience a drastic fall in E′ unlike in the case of BNM. This is because BNMC is integrated into the matrix through chemical bonds. The partial plasticization by the elastomeric additives must be responsible for low ILSS and poor compressive strength of the laminates in these cases.

In the cases of the SPM series of polymers, only a single transition was observed at around 225 °C, which is marginally lower than that of pure BACY. It appears that the two phases are miscible, as otherwise a second transition due to pure BACY should have appeared at higher temperature. The uniform dispersion of the high-T_g polymer and strong matrix/reinforcement interphase aided by the dispersed phenol functions must be conducive for the enhanced mechanical properties for this system.

8
Blends and IPNs of CEs

8.1
Blends with Epoxy and Bismaleimide Resins

Despite their many attractive features, the search for further improvements in performance and reduction in cost of cyanate esters is never ending and a large number of studies have been devoted to matrix modification through blends, and co-curing of these systems. Reactions of cyanate esters with a variety of functional groups like amines [169], hydroxyl [34, 66], epoxy [170–172], phenols, etc., have been reported, among which the most studied are reactions with epoxy [173, 174]. Epoxy-cyanate blends are common and found in many commercial and patented resin formulations.

8.1.1
CE-Epoxy Reaction Mechanism

Although the identification of the co-reaction between the two is as old as the cyanate ester resins, the actual mechanism and the product distribution are still subjects of much debate. Bauer et al. [171, 172, 175–177] proposed a mechanism, according to which, the main reactions occurring during cyanate ester-epoxy curing are (1) trimerization of cyanate, (2) insertion of glycidyl ether into cyanurate, (3) isomerization of alkyl substituted cyanurates to isocyanurates, (4) build-up of oxazolidinone, (5) phenol abstraction, and (6) phenol glycidyl ether addition. These are represented in Scheme 17. The reaction kinetics were modeled by a system of differential equations assuming first-order kinetics. The resulting concentrations of the structural elements were combined with the help of a cascade formalism to describe the statistics of the network formed, whose characteristics could be simulated. The individual rate constants could also be evaluated. Modeling of the reaction kinetics by a system of differential equations was also done [177]. The model was applied to the co-reaction of BACY with 2,2-bis(4-glycidyloxy phenyl) propane to study the gelation behavior and network structure. Up to gel point, trimerization of BACY was the dominant reaction. But the model did not reflect the experimental data at high -OCN conversions. Shimp and Wentworth [170] substantiated the reaction mechanism of cyanate-epoxy reaction proposed by Bauer, using model compounds and isolating the intermediates. The effect of a few catalysts was also examined. Titanium chelates led to more of the toughening and oxazolidinone chain extension products with no penalty in thermal properties. Cyanate ester-diepoxide matrix showed good processability, improved toughness, and hot-wet performance equal to or superior to polyfunctional epoxy-aromatic amine systems. Low dielectric properties

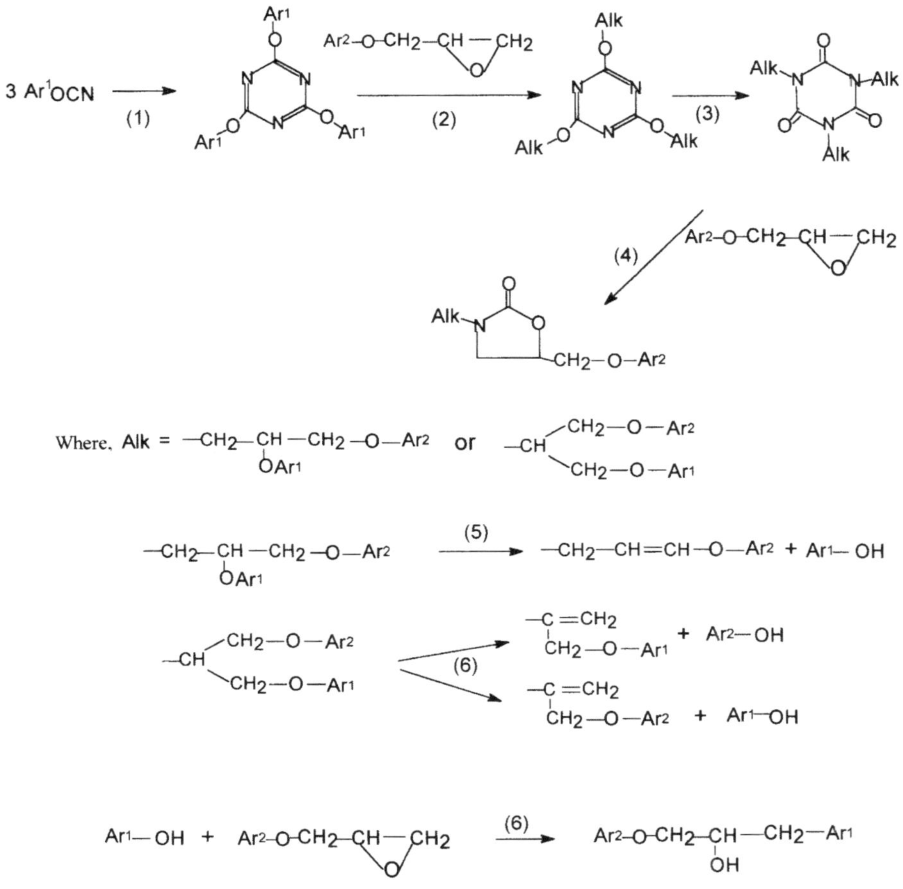

3 Ar^{1}OCN ⟶ (1) ⟶ (2) ⟶ (3)

(4)

Where, Alk = —CH$_2$—CH—CH$_2$—O—Ar2   or   —CH(CH$_2$—O—Ar2)(CH$_2$—O—Ar1)
                              |
                              OAr1

$$-CH_2-CH(OAr_1)-CH_2-O-Ar_2 \xrightarrow{(5)} -CH_2-CH=CH-O-Ar_2 + Ar_1-OH$$

$$-CH(CH_2-O-Ar_2)(CH_2-O-Ar_1) \xrightarrow{(6)} \begin{cases} -C(=CH_2)-CH_2-O-Ar_1 + Ar_2-OH \\ -C(=CH_2)-CH_2-O-Ar_2 + Ar_1-OH \end{cases}$$

$$Ar_1-OH + Ar_2-O-CH_2-CH(-O-)CH_2 \xrightarrow{(6)} Ar_2-O-CH_2-CH(OH)-CH_2-Ar_1$$

Scheme 17. Bauer's mechanism for cyanate –epoxy reaction

of the blends are desirable features for meeting the emerging need for advanced radomes, microwave antennas, and stealth aircraft composites. From uni-directional (UD) composite properties, aircraft service temperature was predicted to be in the range 120–150 °C. Improved alkali resistance of the systems relative to bismaleimides and cyanate ester homopolymer matrices offers superior resistance to galvanic corrosion, where epoxy-amine falls short of thermal requirement or processability.

Using monofunctional model compounds, Fyfe et al. recently studied the reaction by high resolution ^{1}H-, ^{13}C-, and ^{15}N-NMR spectroscopy and mass spectroscopy [178]. The major cross reaction product is a racemic mixture of enantiomers containing an oxazolidinone ring formed from one cyanate and two epoxy molecules. Epoxy consumption lags behind the cyanate consumption as triazine formation is faster than both the self-polymerization of epoxy and cy-

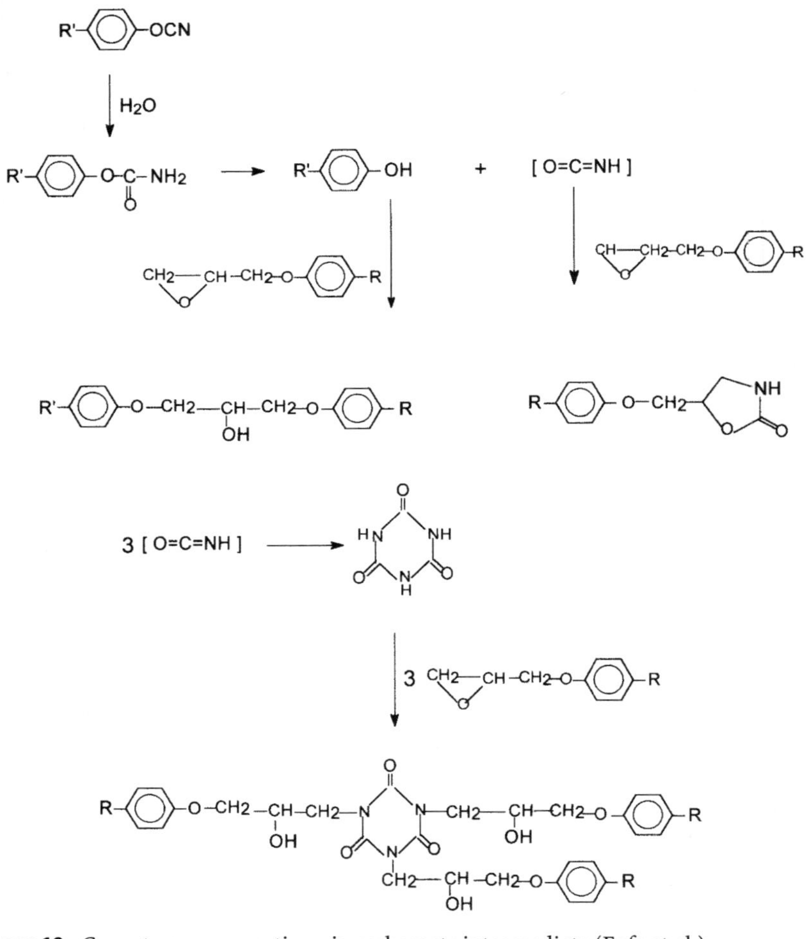

Scheme 18. Cyanate-epoxy reaction via carbamate intermediate (Fyfe et al.)

anate-epoxy co-reaction. The cross reaction between cyanate and epoxy, leading to oxazolidinone is limited to only 12% in view of the higher proportion of epoxy needed for this reaction, in contrast to 32% detected by Shimp and Wentworth [170]. One non-negligible path is the reaction between the carbamate, derived from cyanate and epoxy, giving rise to products including oxazolidinone, cyanurate, and diphenyl glyceryl ether, the products being related to the cross reaction between epoxy and cyanate. The overall reaction is shown in Scheme 18. The cross-reaction mechanism is shown in Scheme 19. The cyanate-epoxy reaction was investigated by Grenier-Loustalot et al., also by FTIR and ^{13}C NMR techniques and using mono functional model compounds [94, 96]. They identified, apart from the oxazolidinone, chemical species like the cyanate dimer and car-

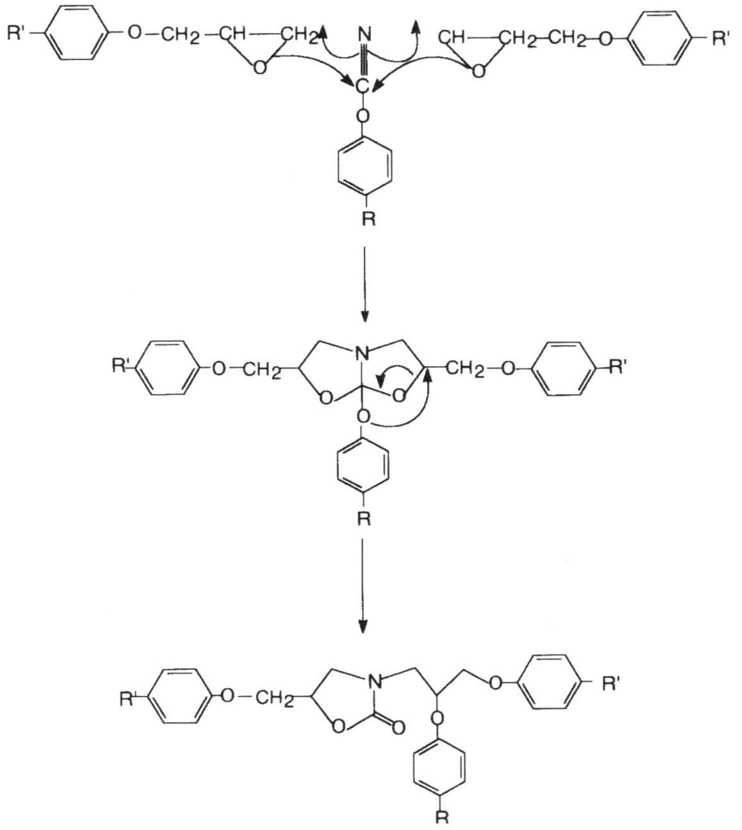

Scheme 19. Cyanate-epoxy cross-reaction mechanism (Fyfe et al.)

bamate type products, which were earlier detected during the homopolymerization of cyanate ester. These products can subsequently react with the epoxy groups. The degradation of carbamate to phenolic derivative was also considered. These data were then applied to the reaction of a difunctional system involving diglycidyl ether of bisphenol A (DGEBA) and bisphenol A dicyanate, in the temperature range 150–210 °C with mixtures prepared in different epoxy/cyanate ratios. It was shown that epoxy functions react on the triazine rings formed in the initial step of homopolymerization of cyanates, and that the structure of the final system depends on their initial relative concentrations. When the epoxy is present in excess, there were no residual cyanate or cyanurate functions at the end of the reaction. A 1:1 epoxy/cyanate ratio was needed for appreciable percentage of cyanurates. The study of the cure reaction in presence of an anionic (imidazole) and metallic [Cu(acetyl acetonate) and Cr(acetyl acetonate)] catalysts showed little change in the reaction path with reference to the non-catalyzed reaction, although the type of catalyst affects the distribution of

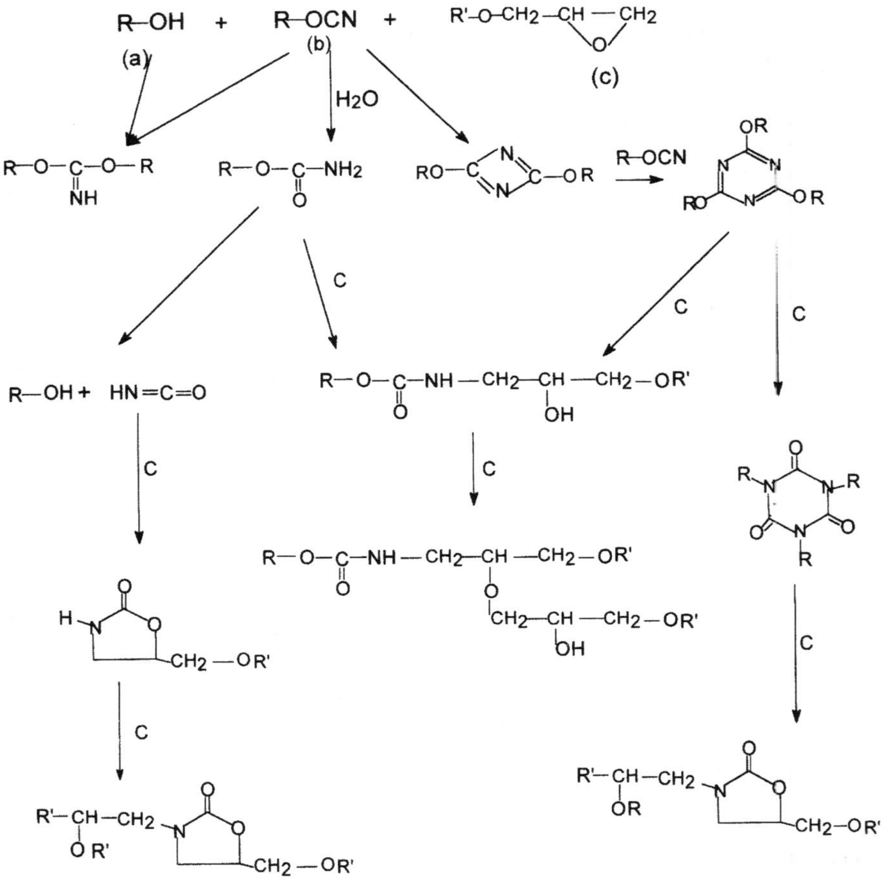

Scheme 20. Cyanate-epoxy co-reaction pathways and products (Loustalot et al.)

the products at the end of the reaction. For example, imidazole led to low concentration of oxazolidinone, and chromium acetyl acetonate did not change the reaction products. Thus, the structure and composition of the network of the final system are complex and nearly unpredictable and depend on the initial concentration of the reactants and the catalysts for a specific monomer. The authors proposed a reaction path accounting for the formation of cyclic dimers, iminocarbonate, carbamate, cyanurate, isocyanurate, and finally oxazolidinone [96] as given in Scheme 20. These findings and the proposed reaction paths agree partly with the finding of Fyfe et al. [178].

A very recent study based on FTIR analysis of the isothermal co-reaction between tetrafunctional epoxy and cyanate ester resins at different stoichiometric ratios and temperatures substantiates some of these findings. Rheological characterization of the curing showed delayed vitrification with increasing epoxy content, indicating the dilution effect of the epoxy resin on the cyclotrimerization of the CE resin [179]. There was little change in the reaction path with reference to the non-catalyzed system when the study was conducted in presence of anionic or metallic catalysts. According to Fainleib et al. [180], who studied the kinetics of the co-cure, the consumption of cyanate ester is rapid at the beginning. They postulated for the first time the formation of oxazoline intermediates apart from cyanurate. At high temperature, the epoxy adds on to the isocyanurate formed in turn from cyanurate, and the oxazoline rearranges to oxazolidinone. In prepregs, the carbon surface modified the cure kinetics of this reaction as evident from FTIR studies [181]. Fyfe et al. also reported A-B type crosslinking monomers possessing both an epoxy and a cyanate group, typically bisphenol A mono cyanate monoglycidyl ether (Scheme 21) which gave tough and strong material on heat curing [64]. Although Scheme 21 indicates straight formation of the desired compound, purification by solvent extraction and column chromatography etc. was needed to eliminate the other statistically possible by-products. Interestingly, the authors claim selective trimerization of the cyanate group, while keeping the epoxy group intact. This was done by heating the monomer to 180 °C to give a tris epoxy cyanurate, a potential epoxy resin, with mass spectral and ^{13}C NMR evidences for its selective formation. It is not very clear why the possible co-reaction between the epoxy and cyanate did not proceed in this case at this temperature. Attempts to polymerize the epoxy group selectively, on the other hand, ended up in reaction of both epoxy and cyanate groups. For example, reaction with diethyl amine in 2:1 cyanate-amine molar ratio at room temperature gave a mixed substituted triazine as shown in the same scheme. Surprisingly, cyanate group was found to be more reactive towards diethylamine than epoxy. This cyanate-epoxy monomer also rendered a cyanate-epoxy blend stronger and tougher. They also reported the crosslinking polymerization of 2-allyl phenyl cyanate by itself, with other dicyanates, and with olefinic monomers.

Cyanate ester has been reported as cure accelerator for epoxy and was also conducive for increasing the T_g and thermal stability of the cured network when diglycidyl ether of bisphenol A was cured by 4,4′-diamino diphenyl sulfone in the presence of BACY. The reaction was followed by FTIR and rheometrically [182]. The cured products from multifunctional, naphthalene-containing epoxy resin and BACY were reported to exhibit better T_g and lower coefficient of thermal expansion in comparison to commercial epoxy systems [183]. The thermal stability, T_g, and moisture absorption of the cured matrix increased on increasing the epoxy functionality. The systems exhibited enhanced thermal stability when cured in presence of BACY compared with when cured in presence of diaminodiphenyl sulfone. Thus, the temperature corresponding to 10% decomposition was higher by 17 °C and peak decomposition temperature higher by 45 °C for the BACY-cured system. Addition of metallic catalysts to the cyanate-epoxy system promoted the cyclotrimerization of cyanate and polyetherification of epoxy. However, the naphthalene-containing system exhibited lower thermal

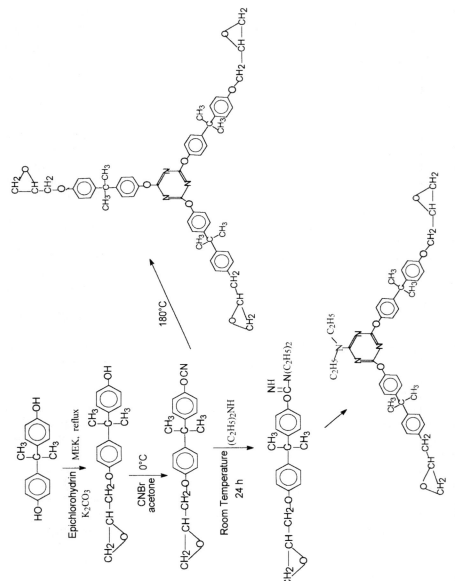

Scheme 21. Synthesis and reactions of A-B type cyanate-epoxy monomer

stability than the bisphenol-based epoxy resin, due to the steric hindrance to crosslinking by the bulky naphthalene rings. Presence of the hydrophobic backbone of the former was responsible for its reduced moisture absorption.

As mentioned earlier, the cyanate-epoxy systems find many commercial applications. Many multifunctional formulations, principally based on CE/epoxy, have been claimed. Generally, the properties of epoxies are improved on co-curing by cyanate ester and the blend is more cost-effective than cyanate alone. The applications of these blends, as covered by patents, has been referred to in a review by Penczek and Kaminska [184]. Most of their applications are in copper clad laminates [185–191], fire resistant formulations [192], aircraft structures [193], and in semiconductor devices [194–196]. Due to the commercial significance of the blends, the information on the epoxy-cyanate systems is still covered by patents and some of the most relevant and recent ones have been referred to in this review.

Dimensionally critical structures like optical support structure of satellites need a minimum coefficient of moisture expansion (CME) property. This property of a proprietary cyanate ester/graphite and cyanate ester-epoxy blend/graphite has been found to be superior to that of the industrial standards based on tetraglycidyl epoxy/graphite systems [197]. Cyanate-epoxy formulations (containing cure catalysts, fillers, and coupling agents, etc.) are reported to have excellent humidity and thermal shock resistance. They find application as underfill materials for flip chip semiconductor devices. The patent also claims good filling, resistance to crack formation on thermal cycling, and good moisture resistance for these formulations [198, 199]. In another patent application, copper clad glass laminates of cyanate ester-epoxy resin varnish (containing cure catalysts, fire proofing agents, and antioxidants) are claimed to possess high T_g, good moisture and electrocorrosion resistance, and good thermal stability. The laminates exhibited UL94 rating V-0 and copper peeling strength 1.6 kN/m [200]. In yet another development, flexible interpenetrating polymer networks of epoxy and cyanate ester resins were prepared using a polyamide catalyst. The system possessed a pot life of 18 h. at ambient and was curable in 1 min at 200 °C. The composition required no solvent for processing and developed high adhesion and high moisture resistance with low mass-loss during cure. This fast curable system is claimed to have application as a die-attach adhesive, heat sink attach adhesive, encapsulant, or underfill for semiconductor assemblies [201].

The blend of the low molar mass novolac cyanate ester prepolymer described above [85] with an epoxy resin (Epikote 1001) was reported to possess good storage stability. The copper clad laminates of the system exhibited high T_g (249 °C), high solder heat resistance, and high copper adhesion strength. The prepregs were stable for 30 days and developed no adhesion with each other on storage at 25 °C and RH of 45%. Gel time at 170 °C decreased from 100 s to 75 s on storage for 180 days. However, when the above cyanate ester prepolymer was blended with the epoxy resin after reacting with bisphenol A dicyanate (1:4 ratio), the copper clad laminate showed a lower T_g (230 °C) due to incorporation of the more flexible bisphenol backbone [202, 203]. The system retained good storage stability for the prepregs, high solder heat resistance, and good adhesion strength. Good heat resistant systems useful in electrical insulators for printed circuit boards are comprised partially of crosslinked product of tetrabromobisphenol A diglycidyl ether and imidazole, polyfunctional cyanate ester, and zinc naphthenate catalyst. The cured systems reportedly possessed T_g of 165 °C, dielectric constant of 2.77, and loss tangent 0.0010 [204]. Formulations, melting at 75 °C, possessing good heat resistance and low thermal expansion, were developed based on BACY and epoxy resins. The compositions are useful as packaging materials for semiconductor devices [205]. Die-attach pastes comprising epoxy resin and cyanate ester components are claimed to possess good wire bonding capability, suitable for binding Si chips onto copper frames [206, 207]. Such compositions also exhibit very low moisture absorption (0.16% without bleeding) [208]. A literature survey shows that most of the published information on CE/epoxy systems refers to their reaction mechanism, kinetics, cure behavior, and thermal stability. Since the co-reaction between the two is complex and is dependent on various parameters, the prediction of properties of the blend becomes difficult, since they do not go by the rule of mixtures. Blends of cyanate esters with bisphenol-based epoxy resins find increasing application in spacecraft structures due to their very low out-gassing property and good dimensional stability [209]. The prospects for enhanced thermal stability for the resultant network increase by choosing a novolac-based epoxy. Several patent applications claim multi-component systems containing BACY, novolac epoxy resin, and other reactants to be having superior thermal stability compared to the conventionally cured bisphenol A-based epoxy or novolac-epoxy resins [138].

In a recent study, reactive blends of BACY and a novolac epoxy resin (EPN) were investigated for their cure behavior and mechanical, thermal, and physical properties of the co-cured neat resin and glass laminate composites [210]. The system with good flow properties manifested a linear relationship between ln (T_{gel}) and 1/T for a 1:1 equivalent mixture. This is shown in Fig. 13. The decrease of gel time with increasing concentration of the epoxy-component was attributed to the promotion of the crosslinking through the multitude of reactions. Contrary to the apparent observation in DSC, the dynamic mechanical analysis confirmed a multi-step cure reaction of the blend, in league with the established reaction path for similar systems. The non-isothermal DMA of the prepreg, which was also used to optimize the processing of the blend, is shown in Fig. 14 for a 1:1 mixture of BACY and EPN. The cured matrix was found to contain both polycyanurate and oxazolidinone networks which existed in discrete phases exhibiting independent glass transitions in DMA. Whereas the low temperature transition corresponded to the oxazolidinone phase, the one around 250 °C was assigned to the residual polycyanurate that was not transformed to the polyoxazolidinone. The DMA of the cured resins are shown in Fig. 15. The flexible and less crosslinked oxazolidinone network contributed to enhanced flexural strength at the cost of the tensile strength of the neat resin. The trend in compositional de-

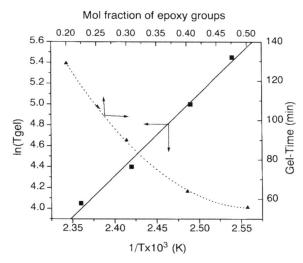

Fig. 13. Dependency of gel time on composition and temperature for the cyanate-epoxy blend

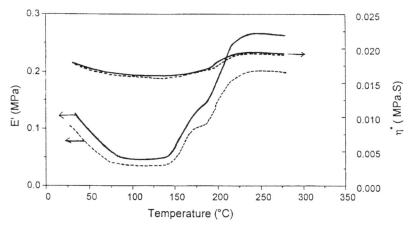

Fig. 14. DMA of prepregs of glass/cyanate-epoxy blend of epoxy-mol % (——): 50%; (---): 30%. Heating rate 5 °C/min, 1 Hz

pendency of mechanical properties of the laminate composites did not bear any direct correlation to that of the neat resin. Although the polar heterocyclic oxazolidinone favored a stronger interphase (and thereby enhanced ILSS), the effect was not reflected in the flexural strength of the composite which showed a decreasing trend with increase in oxazolidinone content. A reverse trend in flexural strength of the neat resin and that of the composite was observed which substantiated the general observation that neat resin properties cannot be extrapo-

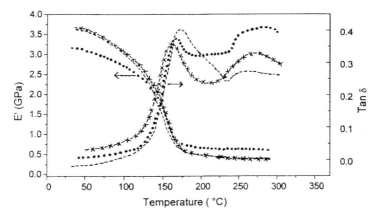

Fig. 15. DMA of cured cyanate-epoxy neat resin blend. Epoxy concentration (mol %) (- - - -): 50%; (-×-×-×-): 30%; (●●●●●●): 20%. Heating rate 5 °C/min, 1 Hz

lated to the composite on a one to one basis. The fact that the increased flexural strength of the neat resins was not translated to the composite implied a complex failure mechanism of the composite under flexural stress and possibility for a different matrix composition in the interphase of the composite in comparison to the bulk. The mechanical properties of the neat resin and composites are given in Table 4. The coefficient of thermal expansion (CTE) values of CE/epoxy resins at higher temperature increased with increasing epoxy content in the blend. The dielectric properties and moisture absorption of the cured blend did not show major deviation from those of the pure polycyanurate contrary to the general trend where presence of epoxy enhances these properties. Table 5 gives the physical properties of the cured blend. The presence of oxazolidinone adversely affected the thermal stability of the cured resin and the high temperature

Table 4. Mechanical properties of cyanate epoxy-blend: neat resin and E-glass laminates [210]

Mol fraction of epoxy groups in the blend	Tensile strength (MPa) for neat resin	(0°) Compressive strength (MPa) for composite	Flexural strength (MPa)		ILSS (MPa)	
			Composite	Neat resin	Ambient Temperature	Retention of ILSS at 150 °C (%)
0.50	33	270	410	114	61	47.5
0.40	24	222	490	99	67	55.2
0.30	25	156	400	95	60	56.7
0.20	37	208	485	–	44	–
0	70	293	501	90	58	65.5

Table 5. Physical and dielectric properties of cyanate-epoxy blends (neat) [210]

Mol fraction of epoxy groups in the blend	Moisture Absorption (%)	Dielectric constant (1 MHz, 60% RH)	Dissipation factor (tan δ)	CTE (10^{-5} °C^{-1})	
				25–70 °C	70–100 °C
0.50	1.53	3.67	0.021	5.25	7.10
0.40	1.46	3.72	0.017	4.85	6.50
0.30	1.47	3.70	0.016	4.90	7.00
0.20	1.54	3.67	0.016	4.85	6.70
0	1.49	3.61	0.022	4.80	6.00

performance of both neat resin and the composites. Co-curing epoxyresin with BACY has been reported to result in enhanced hygrothermal resistance of the blend [211]. The hygrothermal resistance was measured in terms of the change in T_g and mechanical performance of the composites. Immersion in boiling water for 72 h led to moisture absorption of 1.6% only with a concomitant decrease in T_g by about 10 °C. The composite with T-300 carbon fiber absorbed only 0.36% moisture as against the pure epoxy composite absorbing 1%. The composite exhibited excellent retention of mechanical properties after hygrothermal treatment. The authors, who investigated the cure mechanism by FTIR, claimed formation of oxazoline with its absorption at 1695cm^{-1}, contrary to many others claiming oxazolidinone as the product of cyanate-epoxy reaction.

8.1.2
Blends with Bismaleimides

The bismaleimide (BMI) systems dominate over the epoxy-based structural polymer matrix composites owing primarily to their high performance-to-cost ratio and relatively high temperature resistance [212]. However, their processability is difficult and fracture toughness not good. Unmodified BMIs possess high rigidity and brittleness due to high crosslink density, which are major hurdles for material durability. Attempts to reduce brittleness by way of reduction in crosslink density through structural modification, toughening, etc., may adversely affect the high temperature performance. Advanced applications use blends of various bismaleimides and amines to provide a balance of stiffness, toughness, processability, and thermal stability. One such BMI blend formulation known as compimide-353 is included in Scheme 22. Cyanate esters, on the other hand, have very good processability and toughness. However, their thermal stability is marginally inferior to that of bismaleimides. One apparent method to derive both thermal stability and processability is to realize a blend of the two resins. Such blends could possibly benefit from the good physico-chemical

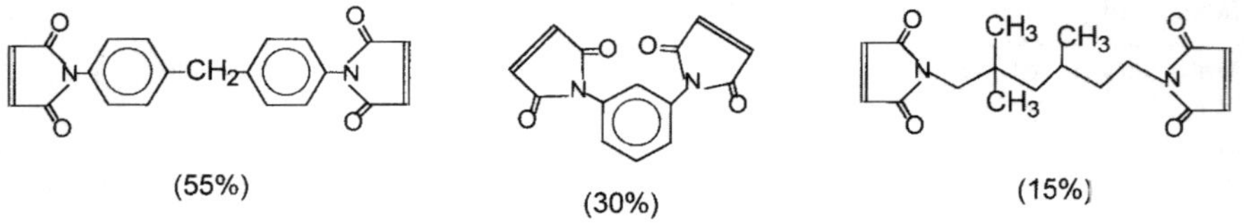

(55%)   (30%)   (15%)

Components of compimide-353 ⊖

2,2- bis (3-allyl-4-cyanatophenyl)propane   2,2- bis (3-propenyl-4-cyanatophenyl)propane

BACP   BPCP

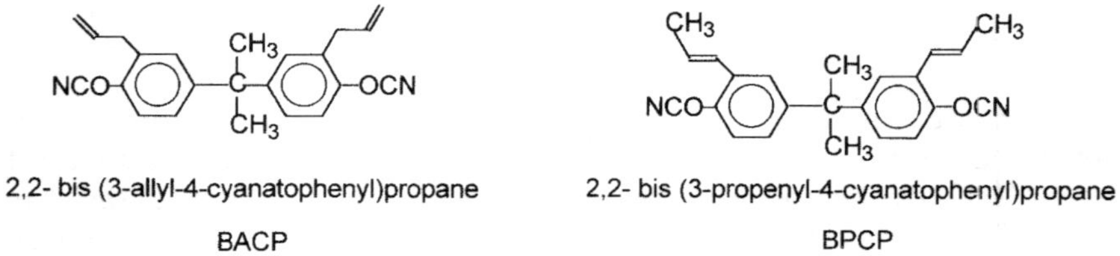

Additives for compimide-353

Scheme 22. Structure of components of compimide-353 and its additives

attributes of the two systems. Blends of BMI and cyanate esters have been quite well explored.

Various researchers [213, 214] have reviewed the reaction of cyanate esters with maleimide. The mechanism is not well established. The earlier postulations were based on a co-reaction between the two and commercial blend formulations of bismaleimide-cyanate known as B-T resins such as Skyflex of Mitsubishi Gas Chemical Co. were introduced [215, 216]. The components of this formulation are BACY and bis(4-maleimido phenyl) methane (BMM). Several patented B-T resins formulations are claimed which find applications as engineering materials in aircraft, reinforced plastics, and injection molding powders, materials in high speed circuits, in electric motor coil windings etc. [109, 184]. With attractive properties such as chemical resistance, dimensional stability, insulation, low dielectric constant, reduced moisture absorption, etc., they are preferred in circuit boards and semiconductor encapsulants. The postulation of co-reaction in such blends did not receive much acceptance in the absence of sufficient proof for the resulting pyrimidine structure [217]. Detailed studies using FTIR and ^{13}C NMR techniques gave no evidence for addition between cyanate and maleimide groups [218]. The reaction between the two, to form pyrimidine structures as originally proposed [219], was examined by Barton et al. [220] using model compounds and with the help of heteronuclear solution ^{15}N NMR. No evidence for any such product was obtained. The components were found to undergo independent polymerization.

Assuming co-reaction, the cure reaction of a mixture of bis(4-maleimido phenyl) methane and BACY was followed by FTIR [221]. The reaction kinetics, studied by DSC, suggested dependency of cure mechanism on blend composition. The apparent activation energy computed by the Prime method increased with BMI content. The rate maximum at a fractional conversion range of 0.32–0.33 indicated an autocatalytic nature of the reaction. The different pattern of activation energy with fractional conversion for two different blend compositions indicated non-identical cure mechanisms for the two compositions. The cyclotrimerization of BACY occurred during the cure of a 1:2 molar ratio of BMI and BACY. Since activation parameters derived from DSC method are generally not consistent, and since the cyanate cure can be catalyzed by impurities present in BMI, which was not taken into consideration, the authors' conclusions on the cure mechanism based on DSC kinetics can be erroneous.

Lack of support for the co-reaction has now led to the conclusion that the two systems form an interpenetrating polymer network (IPN) and a not a co-cured matrix [215, 222]. The IPN formed in these cases have been found to exhibit two T_gs, and the use temperature of the blend is limited by the low T_g component [109, 215, 222]. The two T_gs result from the microphase separation caused by the two incompatible polymers. One way to obviate this is the addition of network linkers to promote the co-reaction of the components and, thereby, to facilitate partial homogenization of the linked B-T IPN matrix with concomitant improvement in T_g [109]. Allyl phenyl cyanates have been quite well explored for serving as network interlinker in these cases. Several allyl functionalized CEs and their co-reaction with bismaleimides forming linked inter penetrating networks (LIPN) of B-T resins have been employed for achieving compatibilization [223, 214]. Maleimides are known to react with allyl derivatives via the Alder-ene reaction [224, 225]. In allyl phenyl cyanate, it is quite likely that the trimerization of the cyanate either precedes or competes with the Alder-ene reaction. The proposed mechanism [214, 225] is shown in Scheme 23. Incorporation via blending of an inherently tough cyanate ester in a bismaleimide matrix significantly improved the performance of the latter. However, the IPN so produced exhibited two T_gs. Addition of small amounts of alkenyl functional cyanate ester could increase the T_g of the network by linking the polycyanurate with the high T_g bismaleimide, while retaining much of the original morphology of the blend and thus also retaining the high fracture toughness [214]. Another study revealed that increasing the allyl functionalized oligomeric cyanate ester component in a commercial B-T blend did not significantly affect the stability of the system [225].

The physical properties of the blends and mechanical/fracture characteristics of their carbon fiber composites in the case of a commercial cyanate/bismaleimide blend could be substantially enhanced by incorporation of propenyl functional cyanate ester which is capable of leading to linked IPNs. Various blends of a commercial BMI mixture (i.e., compimide-353, components shown in Scheme 22), a cyanate ester (AroCy B-30), and a comonomer with allyl group, typically 2,2-bis(3-allyl-4-cyanatophenyl) propane (BACP) or one with propenyl

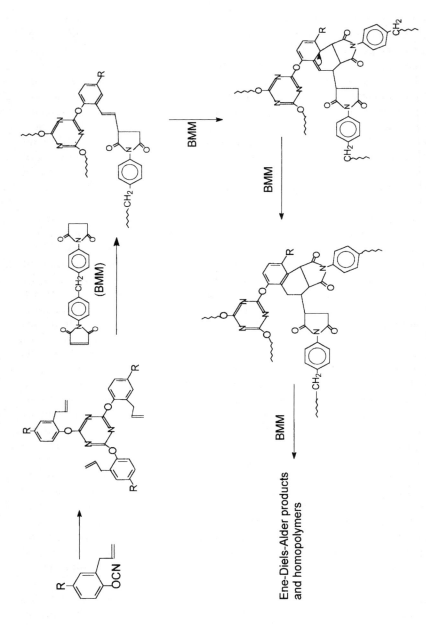

Scheme 23. Reaction Sequences of allyl phenyl cyanale and BMM

function, namely 2,2-bis(3-propenyl-4-cyanatophenyl) propane (BPCP) were prepared by autoclaving [226]. Structures of these additives are also included in Scheme 22. Single T_g value as high as 386 °C could be achieved for such formulations. Interlaminar shear strength of the composites showed an improvement on addition of the propenyl functional cyanate ester to the BMI/CE blend. The blend exhibited improved flexural and compressive properties over the component polymers. However, the three-component blend, incorporating the phase miscibilizer, was detrimental for both flexural strength and modulus. Although the fracture properties showed significant increase in the blend over the component homopolymers, it was achieved at the penalty of T_g. The functionalized cyanate ester was conducive to achieving higher fracture property without penalty over the T_g. The properties of the some of these composites are compiled in Table 6, which shows that the functionalized cyanates are themselves capable of serving as effective matrices. The higher reactivity of the propenyl groups in comparison to the allyl group was believed to be the cause of the better performance of the propenyl-functional cyanate ester in property amelioration of the cyanate/BMI blend [226]. Such linked cyanate/bismaleimide has been found to exhibit a single T_g unlike the commercial blend formulations in other studies. Their thermal and thermo-mechanical properties as studied by TGA and DMA revealed that the thermal stability is comparable or better than that of the commercial cyanate homo- or copolymers [227]. The highly crosslinked networks possessed high T_g and can form compatible blends with flexible monomers to improve processability and fracture toughness. The co-reaction between the two polymers is promoted by the *ene* reaction between the allyl and maleimide groups.

Blends of BMM with modified allyl cyanate alone could also develop useful properties. Such modified cyanate esters include bis{[4-[(3-allyl-4-cyanatophenyl) isopropylidene]phenoxy]phenyl}sulfone and 22-bis[3-allyl-4-cyanatophe-

Table 6. Properties of unidirectional carbon fiber laminates of CE, CE/BMI and linked IPN [226]

Resin	ILSS (MPa)	Flexural Strength (GPa)	Flexural modulus (GPa)	Compr strength (GPa)	G_{IC} (Jm^{-2})
Compimide(CMP, Scheme22)	110.5	11.26	76.3	1.55	176.4
BACY	84.6	1.04	72.4	1.18	285.8
BACY/CMP 1:1	114.5	2.36	99.2	1.33	479.4
BACP (Scheme 22)	105.4	1.44	88.1	1.87	526.6
BPCP (Scheme22)	86.9	1.30	75.7	1.49	363.5
BPCP/CMP/BACY, 1:1:1	100.9	1.63	65.1	1.36	534.2
BPCP/CMP, 1:1	106.4	1.55	53.1	0.91	455.4

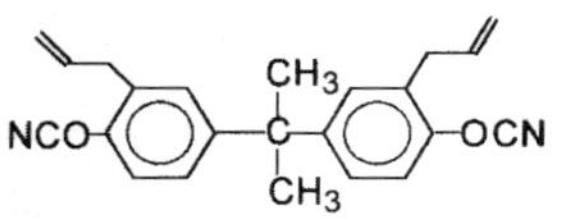

2,2-bis (3-allyl-4-cyanatophenyl) propane 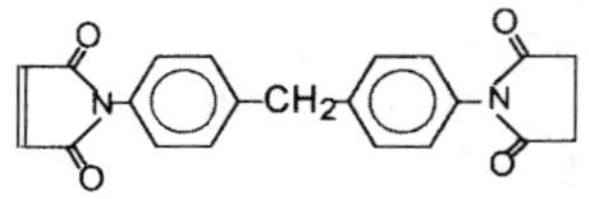   bis(4-maleimido phenyl) methane
BMM

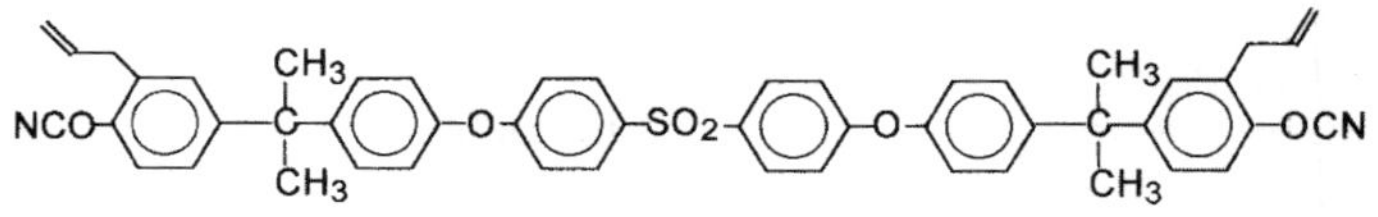

bis{[4-[(3-allyl-4-cyanatophenyl) isopropylidene] phenoxy]phenyl} sulfone

Scheme 24. Structure of bismaleimides and allyl cyanate monomers

nyl]isopropylidene whose structures are shown in Scheme 24. Barton et al. reported a relatively straight and simple route for preparing such alkenyl functionalized cyanate esters, finding potential application as toughening co-curative for bismaleimide and epoxies [228]. A low molar mass additive bearing propenyl group on one end and cyanate on the other is prepared as shown in Scheme 25. Such alkenyl cyanate oligomers, capable of co-curing with BMI resins to form crosslinked networks with improved high temperature properties while retaining the mechanical properties, form the subject of a patent application. They find application in the aeronautical and automotive industries [229].

A recent patent application claims blends of a polyisoimide and a dicyanate ester oligomer in the presence of cobalt naphthenate catalyst which, on curing, yielded a tough thermosetting resin composition with improved low temperature processability. The composition is claimed to possess bending strength of 175 MPa, flexural modulus 3.14 GPa, and good solvent resistance [230]. Cyanate functional maleimide, possessing both functions on the same molecule, is claimed to exhibit good thermo-mechanical properties [231, 232]. A related patent application describes the synthesis of molecules with both cyanate and maleimide functions and their effect on the mechanical properties of a few molding compositions containing BACY or its blend with styrene or ethyl hexyl acrylate etc. [231]. Poly(3-cyanato phenyl maleimide) has better thermal properties than polycyanurates from BACY. Other monomers such as 2-(4-maleimido phenyl) 2-(4-cyanatophenyl) propane, bearing both maleimide and cyanate groups on the same molecule (see Scheme 26) have been found to ameliorate marginally the mechanical characteristics of polycyanurates (from BACY). Thermosetting compositions containing this compound possessed good processability, thermal and mechanical properties, and water resistance [231, 233]. 3-Cyanato phenyl maleimide, cured in the presence of BMM, showed independent curing of each function, the cyanate curing at 93.8 °C (E=44.8±16.8 kJ mol^{-1})

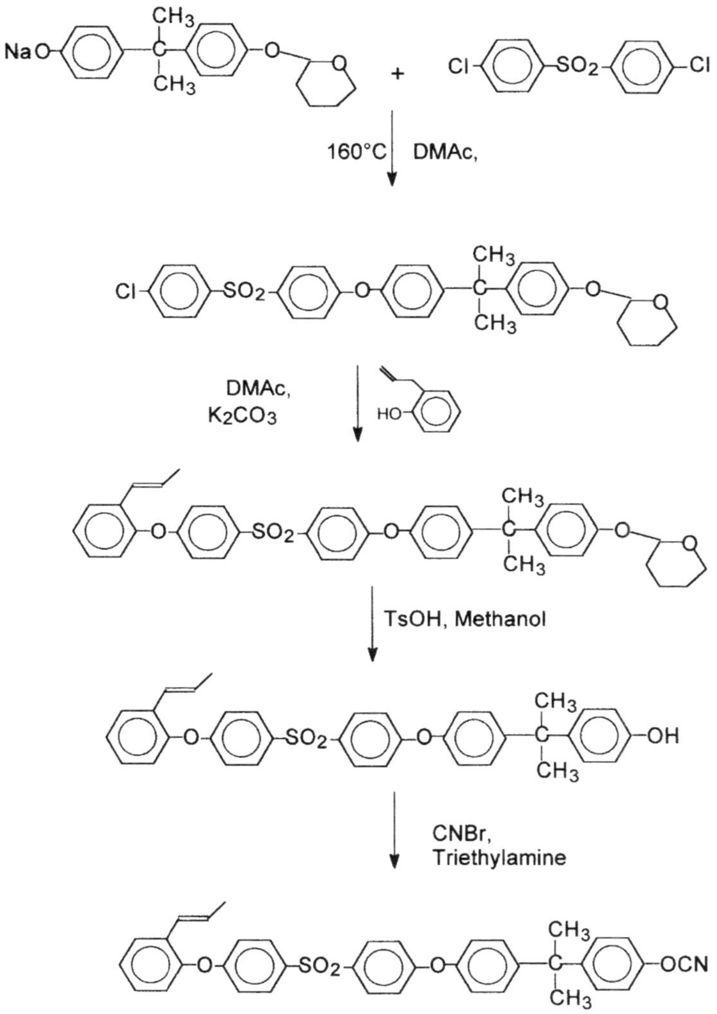

Scheme 25. Synthesis protocol for propenyl cyanate

and maleimide curing at 253 °C (E=81.5±15.6 kJ mol^{-1}) [233]. A recent publication focusing on synthesis, thermal behavior, and properties of naphthalene-containing BT resins points to the structure-property relationship for the cured resins. The T_g, thermal stability, and modulus were higher for higher naphthalene-containing bismaleimide components. Chemical structure of the dicyanate monomers affected the cure behavior and reactivity [234].

If network linking can help make miscible the phases in the IPN of B-T resins, the same could be achievable by using the bismaleimide component having a

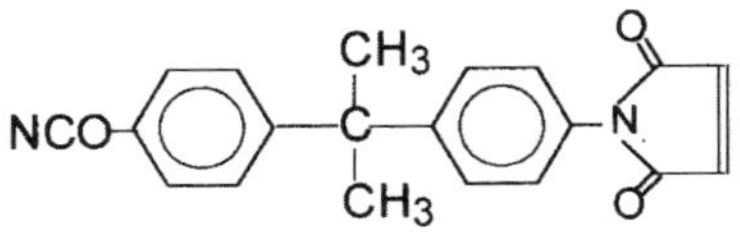

2-(4-maleimidophenyl), 2-(4-cyanatophenyl) propane

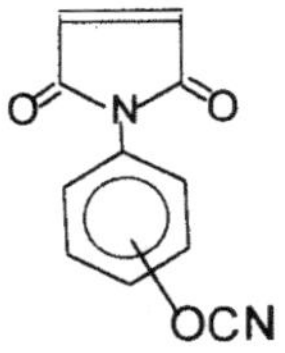

3 (or 4) -cyanatophenyl maleimide (CPM)

Scheme 26. Structure of one- component cyanate-maleimide monomers

structure closely resembling the cyanate ester. This was the basis of a study of the blend of BACY and the bismaleimide, namely 2,2-bis[4-(4-maleimido phenoxy) phenyl] propane (BMIP) where both monomers possess closely resembling backbone structures [217]. The cure characterization of the blends was done by DSC and dynamic mechanical analyses. The near simultaneous cure of the blend could be transformed to a clear sequential one by catalyzing the dicyanate cure to lower temperature using dibutyl tin dilaurate as catalyst. The DSC of the blend and the components are shown in Fig. 16. The broad single cure exotherm of the uncatalyzed blend changed into two independent exotherms on addition of DBTDL, which selectively catalyzes the BACY cure, shifting its exotherm to low temperature regime. This two-stage, independent cure of the components of the blend evident in DSC was confirmed by DMA and led to the conclusion that the copolymerization does not take place, in league with the observations by others. The cure profile of the bismaleimide component predicted from the kinetic data derived from non-isothermal DSC was found to agree with the isothermal DMA behavior. The DMA result of the uncured catalyzed blend, manifesting the two-stage cure, is shown in Fig. 17. The DMA served to optimize the cure schedule as well. The cured blends underwent decomposition in two stages, each corresponding to the polycyanurate and polybismaleimide. Enhancing the bismaleimide component did not alter the initial decomposition temperature, but led to a reduced rate of thermal degradation at higher temperature. Interlinking of the two networks and enhancing the crosslink density through co-reaction of the blend with p-cyanophenyl maleimide (CPM) did not affect the initial decomposition properties but was conducive to increasing the char residue significantly. The thermograms of the blends rich in BACY are

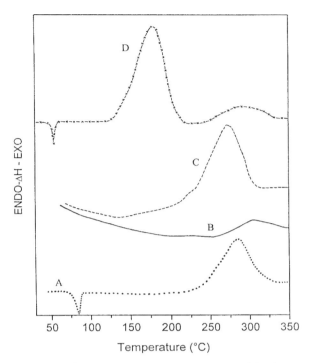

Fig. 16. DSC thermograms of A: BACY; B: BMIP; C: BACY/BMIP blend; D: BACY/BMIP-catalyzed by DBTDL

shown in Fig. 18. Apparently the decomposition of poly BMIP and poly(BACY) occur in a single step in T_gA. A kinetic analysis of the thermal degradation was done using a non-isothermal approach taking into account the major decomposition step evident in the thermogram. The mass-loss range of the degradation step considered for the calculation is given in Table 7. Since the polymer undergoes random decomposition, first order kinetics was assigned without serious error. The kinetic parameters were calculated by plotting $\ln(g\alpha)$ vs $1/T$, where $g(\alpha)$ is a function of α, the conversion as explained in the Coats-Redfern method [131a]. The plots are shown in Fig. 19 in typical cases of the homopolymers which show that the polycyanurate undergoes degradation in a single step, while a two-stage mechanism exists for the bismaleimide for the major degradation step. The kinetic parameters for the two steps have been independently calculated for this polymer.

In the case of the blends, the kinetic plot was linear only for the initial 10–15% mass-loss. Beyond this, interference from the mass-loss arising from the bismaleimide made the linear regression imperfect, showing the existence of a mixed degradation process at higher temperature. Kinetics of the initial degradation step was performed in these cases which showed that the activation ener-

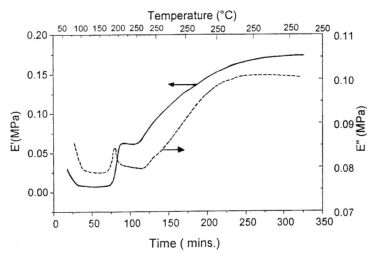

Fig. 17. Cure characterization by DMA (dynamic and isothermal mode) for BACY/BMIP – 60/40 blend at 1 Hz. Heating rate 2 °C/min, in N_2

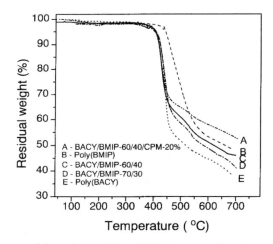

Fig. 18. Thermograms of the poly(BACY) and IPNs in argon. Heating rate:10 °C/min

gy calculated for the initial decomposition of the blends corresponded to the pure polycyanurate. This confirmed that the initial decomposition of the blend is caused by the polycyanurate component. The activation parameters and the degradation steps considered for the computation are given in Table 7.

The structure of the bismaleimide and the cure of the blend are illustrated in Scheme 27. Enhancing the bismaleimide content was conducive to decreasing the tensile properties and improving both the flexural strength and fracture

Table 7. Thermal decomposition characteristics of the homopolymers and BACY/BMIP blends [217]i

Polymer system	Ti (°C)	TIend (°C)	Residue at TIend (%)	Residue at 700 °C (%)	Kinetic parameters[a]		
					Mass-loss range (%)[b]	E (kJ mol^{-1})	A(S^{-1})
Poly (BMIP)	430	570	56.2	48	97–88, step I	209±2	1.5×10^9
					88–61, step II	89.4±2	1.9×10^1
BACY/BMIP-60/40	400	460	65	46	93–83	230±2	2.6×10^{11}
BACY/BMIP-70/30	402	473	61	42	93–82	248±4	1.3×10^{13}
BACY/BMIP-80/20	400	455	63	41	93–84	240±3	1.5×10^{12}
Poly (BACY)	404	458	56	38	93–61	260±1	6.9×10^{13}
BACY/BMIP/CPM	400	460	69	52	–	–	–

Ti=initiation temperature
TIend=Temperature at the end of first stage decomposition
[a] Corresponds to the initial part of the first stage decomposition for the blends and entire part of major decomposition for the homopolymers
[b] Range represents the weight % in thermogram

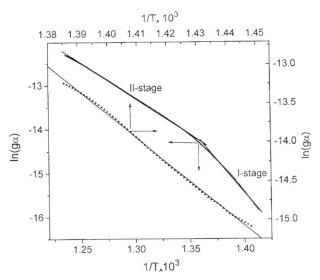

Fig. 19. First order kinetic plot by Coats-Redfern method. ■: poly (BMIP); ▲: poly(BACY)

toughness of the cyanate ester-rich neat resin blend [235]. Typical properties of the blend are given in Table 8.

DMA analyses of the cured blends indicated an apparent homogeneous network for the cyanate ester-dominated compositions. A single T_g was observed

Table 8. Mechanical properties of neat resin moldings of BACY/BMIP blend [235]

Composition (wt.% of BMIP)	K_{1C} (MN m$^{-3/2}$)	Tensile strength (MPa)	Elongation (%)	Tensile modulus (MPa)	Flexural strength (MPa)
0	3.8	70	2.4	3140	95
10	3.5	68	2.3	3100	100
20	3.4	64	2.3	3531	105
30	4.1	59	2.1	3825	114
40	5.3	46	1.6	3603	117

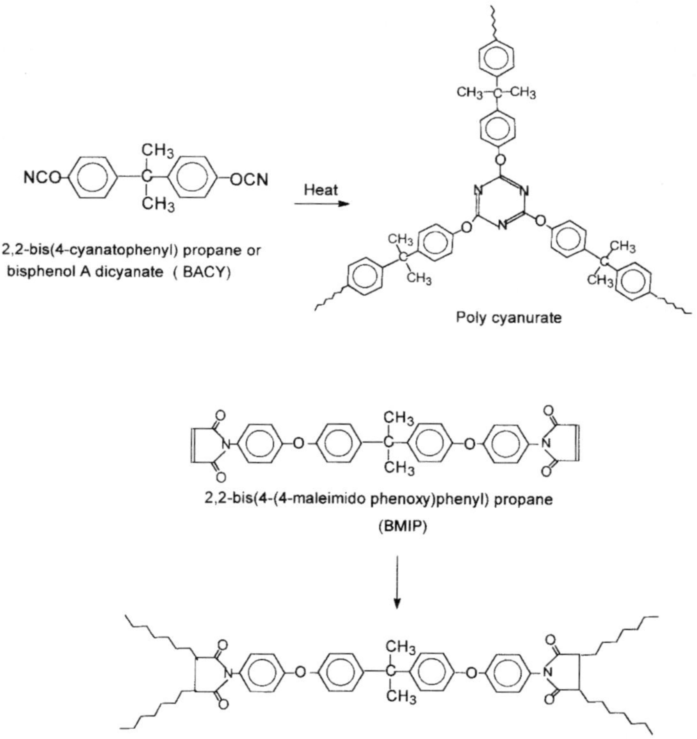

Scheme 27. Strucures of components of IPN and mechanism of network formation (Nair et al. [217, 235])

which remained practically in the vicinity of the T_g of pure polycyanurate as seen from Fig. 20, where an increase in T_g caused by the network interlinker (CPM) can be noted. However, microphase separation was found to occur on enriching the blend with the bismaleimide (beyond 50%). This was evident in the DMA of the cured composites shown in Fig. 21. Addition of bismaleimide did not result in any enhancement in the overall T_g of the blend. Interlinking of the two networks and enhancing crosslink density through co-reaction with 4-cyanatophenyl maleimide impaired both the mechanical and fracture properties of the IPN, although the T_g showed an improvement. Presence of the bismaleimide was conducive to enhancing the mechanical properties such as flexural strength, interlaminar shear strength, and compressive strength of the glass laminate composites of the cyanate ester-rich blend, whereas higher concentration of the imide led to poorer mechanical properties due to brittle interphase. The properties of the composite of the blend are given in Table 9. The similarity in the property-composition profile for both ILSS and flexural strength of the composite as shown in Fig. 22 implied a preferential interlaminar failure in the composites. The IPNs showed reduced moisture absorption and low dielectric constant and dissipation factor, the latter properties being independent of the blend composition (Table 10). However, a recent study on chemical effect of long term water exposure of BMI-BACY blend revealed accelerated water uptake up to the study period of 18 months. Blistering and microcracking were observed, leading to weakening of the blends. The affected blend containing a high proportion of BMI displayed two-phase morphology. Contrary to the virgin material, thermal stability decreased for the BMI-rich aged compositions [236].

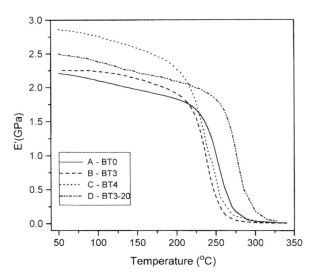

Fig. 20. Dynamic mechanical spectrum of the BACY/BMIP neat resins. 1 Hz. Heating rate 10 °C/min, in N_2. (—)BACY/BMIP-70/30, (····)BACY/BMIP-60/40, (–···–)BACY/BMIP-70/30/CPM-20%

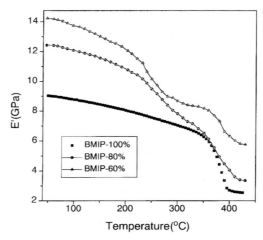

Fig. 21. Dynamic mechanical spectrum of the BACY/BMIP/glass laminate composites. 1 Hz. Heating rate 10 °C/min, in N_2

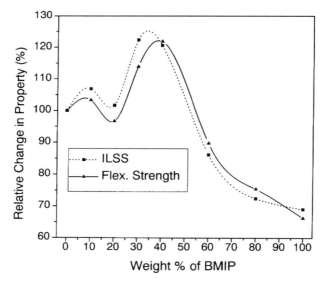

Fig. 22. Dependency of the relative changes in ILSS and flexural strength of the glass laminate composite of BACY/BMIP blend on composition of the matrix

Table 9. Mechanical properties of the glass laminate composites of BACY-BMIP blend [235]

Wt.% of BMIP	ILSS (MPa)	Flexural strength (MPa)	0° Compressive strength (MPa)
0	58	544	293
10	62	562	269
20	59	526	280
30	71	620	307
40	70	663	318
60	50	488	280
80	42	410	250
100	40	340	220

Table 10. Physical properties of IPNs of BACY-BMIP [235]

Wt.% of BMIP	Specific Gravity	Moisture absorption (wt.%)	Dielectric constant (1 MHz)	Dissipation factor (tan δ)
0	1.22	1.49	3.612	0.022
10	1.24	1.34	3.690	0.019
20	1.24	1.30	3.565	0.016
30	1.26	1.27	3.508	0.021
40	1.25	1.25	3.603	0.021

8.2
Reactive Blending with Phenols and Anhydrides

CEs are known to react with phenols to form iminocarbonates which eventually lead to polycyanurates with the liberation of more acidic phenol moiety. This can be a method to alter the gel point of the resin, T_g, and thermal stability of the network by co-curing diphenol with CE. Thus, copolymerization of dicyanate with diphenols resulted in polycyanurates with altered network structure and diminished crosslink density [237]. However, an earlier report claims poly(iminocarbonate) by reaction of these two in equimolar quantities. The thermoplastic so formed was reported to retain the mechanical properties like a polycarbonate. This approach can produce strong, non-toxic, biodegradable films and molded plastics that are degradable at temperatures above 140 °C [169, 238]. Except for a few very early reports [239], the reaction of CE with anhydrides to form poly(iminocarbamates) has not been explored much.

9
Toughening Studies on Cyanate Ester Resins and Composites

Cyanate esters possess good toughness, inherent in the symmetric triazine linked through the ether linkage. However, further improvement in toughness would be demanded for certain critical applications which could be achieved by blending the CEs with various functional and non-functional additives, mainly elastomers and thermoplastics. Such approaches include modification with siloxanes [240], siloxane-polyester block copolymer, amine terminated polybutylene ether [174], amine terminated butadiene-acrylonitrile copolymers [241], rubbers [242], epoxy- and phenol-terminated butadiene-acrylonitrile rubber [243], engineering thermoplastics, etc. Advanced thermoplastics such as polysulfone and polyether ketones have been successfully employed for this [244]. Careful control of heterogeneous morphology is necessary to achieve significant toughening. The solubility parameter and kinetics of phase separation have profound influence on the phase behavior and the mechanical performance. The toughening agents not only enhance toughness but also impart highly desirable solvent stress crack resistance. Thus, hydroxy functional polyarylene ether sulfone rendered cyanate matrix tough without sacrificing the high T_g [245]. The toughenability is dependent on the backbone structure of the thermoplastic. Significant control over the morphology and the toughness is possible through use of reactively terminated advanced thermoplastics. Thus, hydroxy functional reactive polyarylene ether led to controlled macrophase separated morphologies, achieved through systematically varied cure cycles, in contrast to a non-functional polymer of the same backbone and molar mass [246]. Thermoplastic polyether imide has been successfully used for toughening cyanate matrix [247].

Blending cyanate esters with epoxy resins is an accepted method for generating toughened resins without sacrificing hot/wet properties [248]. Epoxy, BMI and reactive rubber can form compatible blends with CEs, modifying the network properties [174]. Blending the proprietary CE resin, XU.71787.02L of Dow Chemical Co. with a proprietary core shell rubber improved the fracture toughness of the matrix and its composite. Incorporation of the rubber neither increased the resin viscosity nor decreased the T_g and shear yield of the host resin significantly. The rubber morphology of the additive was retained and the dispersion remained homogeneous after cure [249]. Cloud-point curves have been determined for blends of a rubber (butadiene-acrylonitrile random copolymer terminated in non-functional groups) and a cyanate ester (1,1-bis-[4-cyanato phenyl] ethane) to understand the phase separation phenomenon. Cloud-points were determined for different rubber fractions in the initial formulation. A thermodynamic analysis based on the Flory-Huggins equation, taking polydispersity of both components into account, led to the following conclusions:

1. Phase separation in the mixture was the result of a decrease in the entropic contribution due to the increase in the oligomer size, and a decrease in the enthalpic contribution as a result of the decrease in the cohesive energy density of the CE-oligomer in the course of polymerization.

2. Spinodal demixing was excluded as a possible mechanism of phase separation during polymerization in solutions containing less than 12% rubber by volume [250].

Pascault et al. studied the blends of two cyanate ester monomers, 1,1-bis(4-cyanato phenyl) ethane or BACY with several initially miscible reinforcing additives as a function of cyanate conversion [251]. The phase diagrams were created. BACY was a better solvent than the other dicyanate ester. Rubber systems based on butadiene-acrylonitrile have an upper critical solubility temperature behavior whereas polyethersulfones (PES) induced a lower critical solubility temperature behavior. The acrylonitrile content of the rubbers and the molar mass of the PES additives also had a great influence on their miscibility. During isothermal cure, phase separation always occurred before vitrification. In rubber it generally occurred before gelation and in PES it occurred together with gelation. The temperature and viscosity at which phase separation occurred were found critical for the final morphology. Reactive additives accelerated the curing process and modified this morphology by inducing a complex matrix-particle interface, and a substructure inside the dispersed particles. These modifications were conducive to the best toughening effects.

Study of the thermal properties of a carbon fiber composite with a cyanate ester, toughened by an epoxy-terminated rubber as matrix, showed no appreciable decrease in thermal properties even at higher elastomer content [252]. Toughening depends on the microphase separated morphology and sizes of the phase separated domains which are also dependent on the backbone structure and terminal groups of the additives. Significant work on cyanate toughening by advanced thermoplastics and related systems has been carried out by Srinivasan and McGrawth. Thus, reactive thermoplastics like OH-functional poly(arylene ether ketone), poly(arylene ether sulfone), poly(arylene ether phosphine oxide), and polyarylene-polyether-polysulfones based on amorphous phenolphthalein were used as reactive additives for various cyanate esters [245, 253–258]. Among these, systems toughened with poly(arylene ether sulfone), poly(arylene ether ketone), and poly(arylene ether) brought about remarkable improvements in toughness and resistance to solvent stress cracking without sacrificing T_g or modulus [256]. Toughening with a high performance thermoplastic polyimide, i.e., Matrimid XU.218, used at several loadings improved the fracture toughness of Primaset PT resin without sacrificing the temperature and mechanical capabilities. A 50% improvement was achieved for 21% loading of the thermoplastic imide. The predominant toughening mechanism was claimed as crack front pinning [259]. Semi-interpenetrating networks with polyetherimide-cyanate ester blends have been reported [260]. Though the macroscopic morphology was unaffected by the cure temperature, the domain size changed. The mechanical and thermal properties underwent drastic change near the phase inversion point. Aryl cyanate ester-siloxane polymers were synthesized as single- or two-phase blends [261]. In the case of two-phase blends, rubbery phase domain size was found suitable for improving the fracture toughness.

Apart from OH-terminated polymers, cyanate-terminated polyether ketone, amorphous phenolphthalein-based poly(arylene ether ketone), poly(ether sulfone), or polyphenylene oxide have also been investigated as reactive toughening agents for cyanate esters [256, 262–264]. In systems toughened with poly(arylene ether ketone) and poly(ether ether ketone), increasing the weight percentage and backbone molar mass of the modifier led to inverted phase morphology at higher weight loadings [256]. However, a miscible blend of bisphenol A-based polysulfone copolymer with AroCy B-10 did not enhance the toughness due to absence of phase separation. For this system, when doped with a non-reactive bisphenol-based polysulfone, morphology and fracture toughness were altered. The significant improvement in fracture toughness, observed relative to reactive bisphenol A-based polysulfone, is likely a result of stabilization between the toughener and matrix [265]. Mechanical behavior and fracture toughness of polymers depend also on the extent of cure and hence the network structure. Thus, fracture toughness of BACY cured to different extents (80–100%) increased with conversion in inverse proportion to the yield stress [152].

Hwang et al. [266] focused their investigations on toughening of cyanate ester with polysulfones. They studied toughening of BACY by polysulfone and cyanate-functional brominated polysulfone. The interfacial adhesion and the domain size changed with the polysulfone content, affecting the fracture toughness and morphology of the blends. Cyanated polysulfone (PS-OCN) formed a phase beyond 20 phr, and particle size of BACY decreased with increase in the polysulfone content. As the optimum concentration for OCN groups in the BACY/PS-OCN blend, 30 phr of polysulfone was fixed. Increasing the interfacial adhesion between the two phases increased the fracture toughness of the clear castings. The increase in interfacial adhesion between the polysulfone matrix and BACY particles was thought responsible for the increase in fracture toughness and the damage zone originated from the ductile yielding of polysulfone component (PSF). The phase separation mechanism and consequent morphology are affected not only by the composition of the blend but also by the curing temperature and the viscosity of the medium [267]. When BACY containing less than 10% polysulfone was cured isothermally, phase separation occurred by nucleation and growth mechanism to form polysulfone particle structure whereas, with greater than 20% of the additive, phase separation occurred by spinodal decomposition to form BACY particle structure. With 15% polysulfone loading, the phase separation took place by a combination of the two mechanisms. The physical and mechanical properties of a polycyanurate containing poly(arylene ether sulfone) in the backbone could be simulated using molecular models [268]. Although the authors claim a close match between experimental and simulated data, a close examination revealed that the simulated properties were systematically higher for the mechanical part and lower for the physical properties. This may be due to the fact that the possible morphological features in case of the poly(arylene ether sulfone)-blended polycyanurate have not been considered for the simulation, since the same method gave a close agreement for the results on unmodified polycyanurates derived from BACY.

Whereas high molar mass polyether sulfones are expected to show phase separation, the low molar mass versions can be miscible and can alter the T_g of the cyanurate network. A recent kinetic study on the reaction of BACY with a low molar mass, cyanated polyether sulfone showed an unexpected S shape for the T_g-composition curve [269]. Studies on model compounds using UV and IR spectroscopy suggested the possible polar interaction between the lone pair of electrons on the N-atoms of the triazine ring and the π^* orbital of the phenylene rings neighboring the sulfone linkage to be responsible for the miscibility and the peculiar shape of the T_g-composition curve.

Brown et al. [270] demonstrated the feasibility of design of custom or gradient morphologies to provide specific mechanical properties of a toughened dicyanate thermosetting resin through intelligent manipulation of the cure cycle and real-time knowledge of the conversion of the system. Fourier transform near infrared spectroscopy using fiber-optic sensors was employed to follow such reactions. Various cure cycle changes resulted in a similar degree of cure, thermal stability, and solvent resistance, but yielded a 20% change in neat resin toughness associated with the morphologies. The morphological variety was shown to occur not only within reasonable cure cycle variations for neat resin, but were also induced through a processing change in a graphite-reinforced composite containing this resin. McGrath et al. [271] explored phenolphthalein-based poly(arylene ether) as toughness modifier for BACY matrix. The additive rendered the system microwave curable due to its higher dielectric loss (than BACY). Controlled morphologies could be generated by varying both the rates of conversion and the composition of the thermoplastic modifier through the use of microwaves. Non-reactive additives led to macrophase separation whereas the reactive ones led to well defined morphologies. The formation of heterophase is strongly influenced by the backbone chemistry and molar mass of the thermoplastic modifier. In another study, BACY was toughened by different ratios of -OH functional phenolphthalein-based amorphous poly(aryl ether sulfone) without significant reduction in mechanical properties [272]. Carbon fiber composites were fabricated by resin transfer molding and resin film infusion with the untoughened and toughened CE systems. The toughened CE/carbon fiber composites possessed significantly improved impact damage resistance compared to hot-melt epoxies. The mode II fracture toughness of the laminates increased with increasing molar mass and concentration of the toughener.

Cyanate-terminated polyethersulfone oligomers can, by themselves, serve as thermosetting systems and have been prepared by end-capping the -OH terminated oligomer with cyanate groups [273]. Carbon fiber composites of cyanate terminated polysulfones [274] or poly (arylene ether sulfones) [275] have been reported. Toughening depended on particle densities in the former case. The laminate G_{IC} (critical strain energy release rate) was improved by up to 80% without significant deterioration in interlaminar shear strength. The compression after impact strength also increased by 7 ksi [276]. Carbon fiber-reinforced composite with an epoxy-toughened cyanate ester/soluble polyimide blend matrix possessed high toughness (with compression after impact strength about twice that of the unmodified cyanate) and 90% retention of the ambient temperature modulus of elasticity at 200 °C [277, 278]. The binary (CE/soluble polyimide) as well as the ternary system (with epoxy) exhibited phase separation via reaction-induced spinodal decomposition. The epoxy content and curing temperature affected fracture toughness of the ternary structure. Thus, addition of the epoxy resin as phase miscibilizer increased the miscibility and reduced the cure temperature.

While thermoplastic additives bring about improvement in toughness, it is usually at the expense of increased dielectric properties which is undesirable for certain critical applications as described earlier. A method to introduce toughness and reduce Dk is based on microporous cyanurates prepared via chemically induced phase separation using a solvent such as cyclohexane. The solvent is removed to generate microvoids by heating the matrix in the vicinity of its T_g. Polycyanurates with significantly less density and D_k were thus prepared [279, 280].

9.1
Interlaminar Toughening

The toughening achieved at the matrix level is not always translated to the composite on a one to one basis. This is because the failure mechanisms are different in both the neat resin and composite. In the case of a tough and strong matrix, the failure in the composite is triggered at the interphase, stressing the need for fortifying the resin/reinforcement interphase for getting good mechanical properties. Addition of the toughener could alter the resin dominant mechanical properties of a composite. Toughening by way of matrix blending is normally achieved at the cost of the thermo-mechanical profile of the composite. One interesting way to toughen a composite without much penalty over the matrix properties is to limit the modification to the interlayer by interlaminar toughening (ILT). ILT has been recognized as a simple means of improving the impact resistance of thermosetting resin matrix, continuous fiber composites, etc. The technique involves incorporating usually heterogeneous organic or inorganic particles, flock, short fibers, or films between prepreg plies. Technology for achieving this varies from simple spray coating of the prepreg with appropriate particles to a complex film-forming process designed to maintain the integrity of the newly interleafed toughening layer during cure. Other methods use the prepreg fabrication stage itself to their advantage by allowing the closely spaced fibers to act as filter, allowing comparatively larger particles on the surface. During curing, the toughened region is formed between the plies of the composites. Cured ILTs manifest exceptionally good impact resistance, reduction in delamination area after impact, and significant increase in mode II fracture toughness. Thus, proprietary reactively terminated PES was used for particulate interlaminar toughening of cyanate ester, BMI, and epoxy composites with carbon fiber [274, 276]. The carbon fiber composites showed an improvement in compression after impact performance. Phase separation leading to particulate toughening (crack stopping or crack front pinning) is thought to be the dominant toughening mechanism. The particles dissolved completely into the matrix during cure, resulting in two toughening effects. At low particle areal densities on the pre-preg, the thermoplastic rich dissolution sites of the particles remained discrete. At higher particle areal densities, these sites overlapped and a more co-continuous film of toughened thermoplastic-rich material was formed at the interlaminar regions. The improvement in G_{IC} was dependent on the nature of the terminal group of PES. In another work, semicrystalline Nylon-6 particulate has been employed for interlaminar toughening of AroCy M-20 resin between plies [163, 166]. The impact properties of the composite improved almost linearly with the particulate doping. The G_{IIC} value of the composite showed a twofold increase over the unmodified system, while the G_{IC} remained unchanged. ILT technique is yet to attain widespread acceptability in aerospace industry.

10
Interpenetrating Polymer Networks (IPN)

IPNs based on cyanate esters are encountered at many phases of blending with thermoplastics and thermosets and do not merit special discussion. Co-curing with bismaleimides results in IPNs and the blends with thermoplastics in the absence of phase separation can be considered as semi-IPNs and are discussed at relevant places in this article. Many IPNs and semi-IPNs have been described in a review article [184]. Simultaneous IPNs were obtained by co-curing BACY with unsaturated polyester. The two polymerizations occurred independently. The T_g and mechanical strength of the polyester were elevated through formation of IPN with polycyanurate [281]. IPN of epoxy-cyanate using polyamide has been described previously. An unusual way of making simultaneous IPN (SIPN) of polyetherimide (PEI) and dicyanate ester has been described by Kim and Sung [282]. The PEI film is inserted into the neat cyanate ester resin and the relative rate of dissolution and diffusion of the former into the cyanate ester is controlled. This is done by controlling the cure of the latter using a catalyst like zinc stearate. This resulted in a morphology spectrum. Due to the concentration gradient of PEI, three types of morphologies were observed. Nodular spinodal structure was formed at concentrations of PEI at around 18%, dual phase morphology having both sea-island region and nodular structure at PEI concentration of 15%, and sea-island morphology for concentrations less than 12%. The system exhibited enhanced tensile and fracture properties, the latter being significantly higher than that of a system with uniform morphology.

11
Liquid Crystalline Cyanate Esters

Cyanate ester-based liquid crystals have interested researchers. The cyanate trimerization at the mesogenic transition temperature can be capitalized to consolidate the formed mesophase. Mormann and Zimmermann have carried out a lot of work in this domain [283–287]. Liquid crystal thermosets, in which the liquid crystalline organization is irreversibly fixed, can be conveniently realized through mesogenic cyanate ester monomers, since the temperature of trimerization reaction can be conveniently tuned by catalyst. The structural changes have a profound influence on the mesophase characteristics [283]. Networks with frozen nematic textures were obtained by thermal cure of (4-cyanatophenyl)-4-cyanatobenzoate and its mixture with 4,4'-dicyanato biphenyl. The isotropic/nematic transition correlated to conversion showed a maximum and minimum, explained by the relative amounts of different triazines formed as a result of the different reactivities of the cyanate groups. On isothermal curing at 160 °C the liquid crystalline phase appeared after 50% conversion as observed by IR, while a critical conversion of 62% was predicted in view of the differing reactivities of the two cyanate functions. The conversion at the gel point was also dependent on cure temperature, varying from 43% at 145 °C to 61.5 at 180 °C [288]. This discrepancy was attributed to the association among the mesogens, decreasing with increasing temperature. The nematic/isotropic transition and T_g were dependent on conversion [284, 285]. The structures of the liquid crystals can be found in Scheme 28. Nematic mesophase liquid crystals were obtained from cyclotrimerization of cyanate-terminated nematic chains [283, 287, 289]. The liquid crystalline properties, mesophase formation, and influence of functional groups on thermal transitions were studied for cyanate- and isocyanate-functionalized compounds [290]. The clearing point was lower by 30 °C for the cyanate-functional system [285, 286]. Ou et al. did the synthesis and polymerization of liquid crystalline aromatic monocyanate (M) and dicyanates (D$_1$ and D$_2$) with a *trans* stilbene structure. The monocyanate on curing gave trimer with discotic properties, exhibiting Schlieren textures at 178–214 °C. Mesophase behavior of the dicyanate ester depended on its structure. Thus, a chloro-substituted derivative formed no mesophase, whereas the non-chloro-substituted one exhibited Schlieren texture above 160 °C. However, the texture froze at higher temperature due to concomitant curing, and a liquid crystalline thermoset with clearing temperature of 290 °C was obtained [291] (structures are included in Scheme 28). Properties of the resulting network were found to vary with the degree of polymerization of the mesogenic polymer. Wang et al. [292] reported liquid crystalline Schiff's bases, terminated with cyanate functions. Curing gave polycyanurates with discotic properties. Cyanate curing is also exploited to stabilize nonlinear optical materials that need poling for activity. Nonlinear optical materials can be incorporated in CE matrix which is cured under poling conditions without degradation [293].

Curing of liquid crystalline cyanate ester resins in electric fields is a new trend in thermoset design and processing and can be used to control directly their mechanical and physical properties. Combining new LC materials with non-LC cyanate monomers leads to a variety of novel ordered network structures and is a convenient method for modifying and controlling their chemical and physical properties [294].

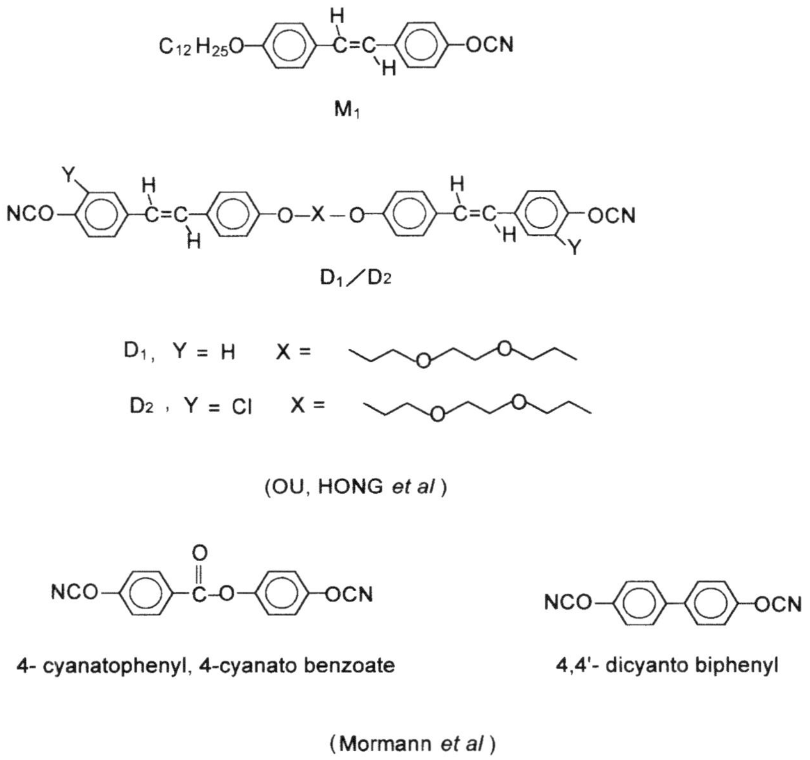

M₁

D₁/D₂

D₁, Y = H   X =

D₂ , Y = Cl   X =

(OU, HONG *et al*)

4- cyanatophenyl, 4-cyanato benzoate   4,4'- dicyanto biphenyl

(Mormann *et al*)

Scheme 28 . Liquid crystalline cyanate esters

12
Degradation Studies on Cyanate Esters and Composites

High performance composites for military aircraft are expected to perform at temperatures in excess of 300 °C and harsh environmental conditions. The limiting factor governing a composite performance is the thermal and environmental stability of the polymer matrix. Among high temperature resistant polymer matrices identified for such a task, polycyanurates also find a place. Polycyanurates are, by and large, resistant to aggressive environments, solvents, acids, UV, and other radiation. Although cyanate ester systems absorb significantly less moisture than epoxies [65, 66], they are prone to undergo hydrolysis-mediated degradation. Studies have shown the presence of phenols, carbon dioxide, and cyanuric acid as hydrolysis products and, accordingly, the reaction sequence shown in Scheme 29 has been proposed for the degradation [295]. In another investigation [296], the hydrolysis of a cyanate ester network made from BACY

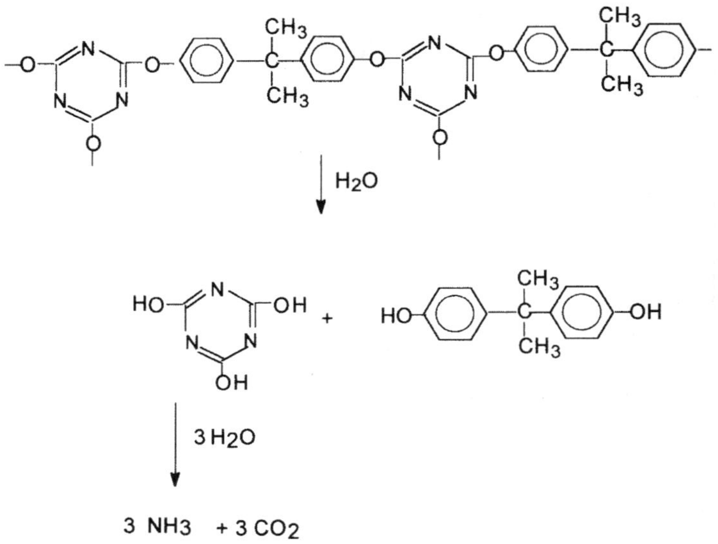

Scheme 29. Products of hydrolysis of poly (BACY)

was performed. Hydrolysis reactions were done isothermally at temperatures from 150 °C to 180 °C under conditions of excess water. The kinetics of the reaction was characterized by the decrease in T_g as measured by differential scanning calorimetry. The rate of change of T_g was found to be adequately described as first-order in T_g, which is an indirect measure of the concentration of crosslink junctions. The activation energy of the reaction was found to be 115 kJ mol$^-$. In addition, moisture-conditioned, glass-reinforced laminate samples were heated and the time for delamination or blistering was recorded as a function of temperature. The blister time at solder temperatures (220–260 °C) was modeled using the above-mentioned kinetic results. Heat transfer to the laminate was considered and the criterion used for blister time was the time at which temperature corresponded to the T_g of the sample. At lower temperatures (<220 °C), loss of water from the laminate was observed to be sufficiently fast to prevent blistering.

Composites of CE systems are more vulnerable to solvent attacks, which affect the interphase. Several studies have been conducted on carbon fiber/cyanate ester laminates related to their thermal cycling tolerance, moisture absorption, and out-gassing, effect of physical aging on mechanical properties, dimensional changes under various conditions, etc. [297]. The effect of co-catalyst, nonylphenol, and copper naphthenate on the thermal degradation of polycyanurate was examined by monitoring T_g and mass-loss. T_g decreased with time and temperature of systems containing the catalyst, and the degradation onset occurred sooner with catalyst concentration. Nonylphenol had no influence. The activation energy was estimated as 50 kcal mol^{-1} from the TGA analysis [298]. Chung and Seferis developed models to analyze the long-term properties of CE/carbon fiber composites in an accelerated aging environment by making use of the change in viscoelastic properties [299]. Dynamic mechanical analysis was employed to monitor quantitatively the aging in an isothermal environment. However, prolonged exposure resulted in extensive matrix microcracking and moisture gain, leading to interface delamination and translaminar cracking in the ply. Thermal-oxidation effects also play a significant role at peak thermal cycling temperature [300, 301]. The glass transition temperature drops during the last step of moisture absorption [157].

Moisture effect on hydrolysis of CEs has been studied in the presence of zinc octoate and copper- and cobalt naphthenates using model compounds. The latter showed ten times less carbamate formation (due to hydrolysis) for AroCy B-40-aramid fabric composites [302]. Different coordination metal catalysts accelerated the hydrolysis of CE-aramid composites to various extents [124]. Influence of various solvents like hydraulic oils, salt spray, acids, and alkalies, as well as the effects of time, temperature, vacuum, and moisture environments on dimensional stability and mechanical properties have been studied. Only a small amount of the environmental liquid was absorbed by the quartz fabric-reinforced CE resin matrix composites on exposure to the various environments. However, they suffered a deterioration in mechanical properties [303, 304]. Physical and chemical aging behavior of CE laminates was evaluated using GC-MS, TGA, NMR, and DMA [305]. The impact of physical aging on damage tolerance and impact damage resistance of graphite fiber/thermo plastic-toughened cyanate esters was evaluated by C-Scan and X-radiographic techniques. Damage increased progressively with aging time and was more drastic in the case of air aging. Tensile strength varied, depending on aging time, environment, and impact velocity [306]. T_g also decreased with aging time [307]. The effect of physical aging on the viscoelastic creep properties of a thermoplastic-toughened cyanate ester resin (Fiberite 954–2) and its IM8/954-2 composites, and a semi-crystalline thermoplastic (Fiberite ITX) and its IM8/ITX composites was investigated using effective time theory, and master curves were drawn. The study was carried out by using dynamic mechanical analysis and tensile creep tests. The tests were performed on plain resin, 90°, and 45° composite specimens. Creep tests were conducted up to an aging time of 54 h with a logarithmic aging shift rate, and its dependence on sub-glass transition aging temperature was determined. The results showed significant physical aging in both material systems. To study the effect of long-term aging on creep behavior, momentary creep tests were conducted on the 45° composites of both material systems at temperatures between 140 °C and 200 °C. Master curve plots were drawn from these momentary creep tests using the time/temperature superposition principle (TTSP). Effective time theory was then used to modify TTSP by incorporating physical aging effects. As aging time increased, the creep compliance decreased consistently. Analysis indicated that a decade of aging corresponded to a 4–5 °C decrease in the test temperature of the system [308]. As a result of physical aging, a decrease in free volume and a corresponding decrease in T_g have been observed [305]. A residual stress measuring instrument was devised and the residual stresses in K135-2u/954-3 carbon fiber-cyanate ester composites were subjected to various stress levels and thermal and hydrolytic cycles were tested [309]. Hydrolysis studies on phenolic triazine (PT) in comparison to an addition polyimide, both potential matrices in high performance composites for military aircraft, revealed that the cyanate system absorbed about 2.1% moisture vis-à-vis the polyimide absorbing around 1% under saturation conditions. This was attributed to the incomplete cure of the cyanate system. The T_g dropped by about 45 °C at moisture saturation level for cyanate as against only 25 °C for the polyimide due to plasticization and possible hydrolysis. Interestingly, in both cases the T_gs were not only completely recoverable on heat treating the hydrolyzed polymer, but that the recovered products showed higher T_gs due to some post curing effect [310].

Although many reports discuss the thermal stability of cyanate esters and their blends, there has not been much study to understand the mechanism of thermal degradation of cyanate esters. Study of the decomposition products of polycyanurates and model compounds has led to the proposal of a thermal degradation mechanism, triggered by hydrolytic cleavage of the ester linkage accompanied by subsequent decomposition of the triazine ring via hetero and homolytic decomposition reactions. The degradation products of cyanurates have been identified as mainly carbon monoxide, carbon dioxide, and hydrogen [311, 312]. The swelling and blistering of cyanurate system encountered at elevated temperature have been attributed to the evolution of carbon dioxide arising from the decomposition of carbamate species formed during the curing of the matrix [313].

Table 11. Comparison between predicted and experimental properties

Property	BACY		PES-CY	
	Predicted	Experimental	Predicted	Experimental
Bulk modulus (GPa)	3.79±0.28	3.89±2.04	2.53±0.28	3.68±2.43
Poisson's ratio	0.35±0.01	0.39±0.13	0.33±0.02	0.35±0.19
Young's modulus (GPa)	3.39±0.16	4.04±2.14	2.50±0.04	3.64±1.86
Lame constant	2.95±0.25	4.32±2.17	1.91±0.29	3.45±2.54
Shear modulus (GPa)	1.25±0.05	1.28±0.75	0.89±0.10	1.12±0.79
Linear CTE (K^{-1})	-	-	1.03×10^{-4}	1.92×10^{-4}
Volume CTE (K^{-1})	-	-	3.09×10^{-4}	5.76×10^{-4}
T_g (°C)	-	-	217	140

13
Molecular Simulation

Hamerton and coworkers have succeeded in simulating the physical and mechanical properties of polycyanurate systems [268, 268a]. Elastic moduli and T_g of BACY and polyarylene ether dicyanates were predicted with reasonable agreement with experimental values. The simulated and predicted properties for both polymers are given in Table 11 . The large variation in some of the properties like T_g is attributed to the wrong assumptions being made, namely that the crosslinking takes place exclusively by trimerization, whereas evidence has been presented for bicyclophane-type ring structures.

14
Applications of Cyanate Esters

The attractive physical, mechanical, and electrical properties of cyanate esters render them the resin of choice in advanced composites for structural and non-structural applications and also in microelectronics.

14.1
Electronics Applications

Polymer materials required for high-speed circuits and microwave transparent structures such as radomes are to be characterized by low dielectric constant and dissipation factor (tanδ) for better electrical performance, signal speed, and low power loss. Moreover, the moisture absorptivity has also to be minimum, as this can alter the D_k and D_f values. These features make cyanate ester the resin of choice in high speed printed circuit boards in the microelectronics industry, where its estimated use is to the tune of 70% [66]. The good adhesive characteristics make them well-suited for applications as interlayer dielectric in integrated circuits. AroCy L-10 and blends with RTX-366 have been used to formulate thixotropic encapsulants for bare mounted chips. Flexibilization using polysiloxane confers good thermal shock resistance [240]. The single largest use for CEs is the lamination of substrates for printed circuits and their assembly via prepreg adhesives into high density, high speed multi-layer boards [314]. The multi-layers are produced commercially for supercomputers, mainframes, and high-speed workstation mother units. High frequency circuits designed up to 12 GHz for wireless communication and tracking systems are the fastest growing use. These fast growing applications include dispatcher radio, pagers, cellular phones, global positioning, satellite broadcast, and radar tracking systems. The demand for cyanate esters in printed circuit boards is likely to increase in view of their low dielectric properties. IBM has introduced a proprietary thermoplastic toughened cyanate ester-based material for high speed electronics [315, 316]. The base resin is a blend of fluorinated and a non-fluorinated cyanate esters. The low loss tangent of such systems is conducive to a high rate of data transmission, important for digital communication applications. The search for still lower dielectric constant CEs like the fluoroalkyl cyanates described previously is ongoing. A few recent patent applications in this context also merit citation [204, 317]. Thus a varnish composition containing AroCy B-10, alkyl phenols, and other components along with silica filler applied on a PET film provided a polymer film with dielectric constant 2.7 (at 1 GHz), loss tangent 0.0044, T_g=172 °C, and tensile modulus 4740 MPa [317]. Other applications of cyanate esters in microelectronics include the use of filled, low viscosity monomers as die attach adhesives in flexible circuits as mentioned earlier [198]. Cyanate esters are likely to find applications in photonics, as optical wave guides, and in nonlinear optics [89, 318]. It is also relevant to cite related recent patents on the development of cyanate ester systems for electronic and low dielectric semiconductor devices [41, 316–338].

14.2
Aerospace Applications

CE systems possess several attractive features required of a resin for application in aerospace. Most of these derive from the unique backbone structure, tailorable to specific applications through structural modification or through blending. The compatibility with various reinforcements and ease of processability by any of the conventional methods are added advantages. However, the high cost in comparison to the state-of-the-art resins limits their applications mainly to the high-value technology areas like aerospace, where cost is only a secondary concern.

They find application in primary and secondary structures in military aircraft. The excellent dielectric properties render them the material of choice in radome applications. The low out-gassing, microcracking properties, and resistance to ionizing radiation and thermal cycling make them suited for satellite applications. The low out-gassing, minimal dimensional changes during thermal cycling, good long term stability, self adherent properties to honeycomb and foam cores, good electrical properties, and high service temperature are the key advantages of cyanate esters over state-of-the-art epoxy resins [334–337]. As a result, cyanate esters are bound to replace the latter in the near future. As the dimensional stability requirements for antennas, reflectors, and optical constructions become more exacting, cyanate esters are projected as the most promising candidates for aeronautics and aerospace applications [338]. High modulus carbon fiber/CE laminates are increasingly used as replacements for Al, Be, and Ti metals in high precision detectors [339]. Cyanate ester-graphite fiber laminates have been used in spacecraft structural equipment, multifunctional satellite bus structures [197, 209], space optical instrument pipes [340], high precision solar-B optical telescope [209], solar array substrates [334, 341], high temperature, high pressure flare housing, shroud, nozzles [342], etc. Ultra-high modulus/high thermal conductivity P130/K1100 pitch-based graphite fiber-cyanate ester composites have been used in the nine Intelsat satellites made by USA in 1993 [338].

For faster and cheaper space programs, minimizing payload mass and volume are critical, particularly for budget-constrained missions. Thus, major cost-saving small satellite programs use K1100 graphite/CE composites for structural design of full capability satellites for the Navy Geostat Follow-on Program [GFO]. The structures met strength, thermal, and electrical requirements [343]. Toughened cyanate ester with very low cure shrinkage, low dielectric constant and moisture absorption, ideal for use ranging from radomes to primary aerospace structures has been developed by ICI Fiberite [344]. One family of CEs, namely 954-1, is designed for use in radomes operating at high frequencies and temperatures. It has controlled flow properties tailored for honeycomb fabrication. Fiberite 954-2 is designed for primary structural applications with enhanced damage tolerance, hot/wet performance, and dielectric properties. The system 954-3 absorbs significantly lower moisture and is resistant to microcracking. The chemical structures of these formulations have, however, not been disclosed [344]. Comparable or better properties for carbon fiber composites of novolac cyanate resins in comparison to PMR-type addition polyimides have been claimed, implying potentiality in structural composite applications [78, 158]. Space durable mirrors with good reflectivity, adhesion, and environmental durability could be obtained due to the space qualities of CEs as discussed above [345]. Feasibility of utilizing them for cryogenic application has also been investigated [346]. CE/AS4 carbon fiber composites have superior interfacial shear strengths (59 MPa) than the corresponding aromatic amine-epoxy matrix-based composite. This is attributed to the better H-bonding strength of the CE system [347, 348].

When dimensional stability of the space structures is a requisite, control of CTE and CME is critical. Near-zero CTE structures are obtainable through choice of fiber, its orientation, and resin content, and hygral expansion is largely dictated by the identity of the resin. Design and fabrication of dimensionally stable platforms, optical benches, and reflectors are greatly simplified by using a CE resin with minimal CME. The absorbed moisture could result in swelling of the structure during fabrication and subsequent desorption in space leads to shrinkage. Dimensionally critical structures like optical support structures of satellites require minimum coefficient of moisture expansion (CME) properties. Figure 23 illustrates the comparative moisture absorption and resulting swelling properties of potential aerospace composite matrices. CE, absorbing minimal moisture, exhibits only one-quarter of the hygrostrain of epoxies and BMIs. The approximate Dk ranges of the various systems shown in the same figure confirm the superiority of the CE resins. Dk values increase with moisture absorption and hence the low moisture absorptivity of CE has a double advantage.

This property, measured for proprietary cyanate esters (Hercules HX-1562, 1565, 1939-3 and 954-2 resins)/P75 graphite combination and cyanate ester-epoxy blend/graphite combination, showed their superiority over the industrial standards based on tetraglycidyl epoxy/graphite systems [197]. Mechanical properties also confirmed that they are suited for dimensionally stable structures. Advanced composite structures capable of withstanding exposure to tem-

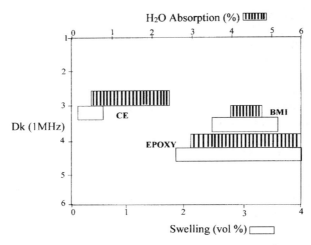

Fig. 23. Comparative hygrostrain behavior and dielectric properties (in non-swollen state) of epoxy, BMI, and CE resins

peratures above 200 °C are required in supersonic spacecraft. These are often realized with bismaleimide matrix. Since these structures have complex shapes and curved surfaces which cannot be repaired using staged prepreg patches or bolt-on metal plates, a wet lay-up patch based on AroCy L-10 cyanate ester has been recommended for highly conformable repair. The low viscosity of the system was advantageous for the wet lay-up. A double vacuum procedure results in void-free laminate with good ILSS. However, its blend with BACY was not fruitful in yielding a good laminate and mechanical strength. Such a patch meets the temperature requirement normally not met by the epoxies [349].

New radomes designed to be used with transmitters and receivers operating at high frequencies (9–44 GHz) ideally require a structure transmitting 100% of the electrical signals. The ideal matrix should possess low D_k and dissipation factor (D_f), low moisture absorption (since this can increase both properties) and desirably high T_g for withstanding the heat generated in case of absorption of the electrical signals. Speciality CE systems with hydrophobic structure and high T_g are ideally suited as matrix for this application. Such resin systems are also qualified to make antennas, missile nose cones, and related applications where electrical transparency and structural performance are required. Composites serving as protective windows or antennas for microwave communication and tracking devices are to be transparent to the passage of microwave. Use of CEs with good dielectric properties ensures reduced reflectance, increased range or signal strength over broader bandwidths, and improved signal quality. Cyanate-based composite structures are preferred to the epoxy-based ones for advanced space structures for reasons of better microcrack resistance, especially on modification by minor amounts of thermoplastics, low moisture absorption typically less than 1% (less than 0.1% condensable volatiles), and resistance to ionizing radiation. Space structures for low earth orbit satellites can be protected against oxygen plasma by using siloxane cyanate polymers which also serves as a toughening agent [240]. Currently, CE composites reinforced with high modulus carbon fibers are preferred materials of construction for advanced communication satellite housing, parabolic antennas, solar panels, precision segmented reflectors, optical benches, struts. and trusses.

Although direct application in aerospace has not been discussed, the development studies on certain systems imply their potentiality for application to aerospace structures. Laminates fabricated from low viscosity cyanate resin AroCy L-10 exhibited low dielectric-loss properties [105]. Damage tolerant composites with negligible moisture plasticization, cure shrinkage, and dimensional change were prepared from the CE resin RTX-366, introduced by Hi-Tek [350]. The resin is claimed to be compatible with a variety of reinforcements like glass, silica, quartz, carbon, polyaramid, etc. Cyanate resins HX-1553 and HX-1562 with undeclared structure have been fabricated into composites with low moisture absorption, good damage tolerance, and good mechanical properties. Filament wound structures from PT resins and carbon fibers reportedly exhibited excellent property retention at higher temperature [78, 158]. Flaw-free, thick walled, hoop wound composite rings with CE/graphite fiber survived high pressure testing, recording a nominal transverse compressive strength greater than 275 MPa [351]. Carbon fiber/semicarbon composites could be prepared by impregnating the C-fiber reinforced C-composite with a heat hardened, cyclized, and dehydrogenated cyanate ester at 150–400 °C followed by heating at 650–900 °C for semicarbonizing the matrix [352]. C/C composites with minimal stress build-up were prepared by sizing the carbon fiber with a BT copolymer or using the copolymer as the impregnating resin [353].

14.3
Adhesion Applications

Despite the absence of polar groups, CEs exhibit excellent adhesion to metals, glass, and carbon substrates. Cyanate esters are extensively used in various formulations for adhesion [354–361], structural film adhesives for microwave applications [314] etc. The adhesion properties of the cyanate systems have been discussed by Shimp [314, 355]. It is believed that relatively low shrinkage, formation of covalent bonds with OH-containing substrates and coordination with metals are the possible adhesion promoting factors. BACY and its blend with epoxies exhibit excellent lap shear strength values, almost double those of the epoxies, that are retained over a wide temperature range up to and above 200 °C. CEs show excellent peel strength to metal foils in comparison to bismaleimides during PCB fabrication. Structural film adhesives based on a cyanate ester (EX-1516) exhibited better retention of ambient temperature adhesive strength at elevated temperature (121 °C), compared to the epoxy-based EX-1548 resin [362]. Silver-filled cyanate ester adhesives for packaging advanced microprocessor chips have also been reported [363, 359].

The study of the interaction mechanism of thin films of BACY prepolymer on different surface states of Si and oxidized Al employing advanced techniques such as XPS, UPS, MIES, IR reflection spectroscopy, and AFM was undertaken by Possart and Dieckhoff [364]. The trioxy triazine was the only moiety identified to have adhesive interaction with the substrate. On a Si surface, the mechanism was identified as donor-acceptor interaction where the lone pairs of electrons on N and O atoms of trioxy triazine were involved in the electron donor process for the Si cation. On aluminum oxide, the Lewis acidic OH groups act as electron acceptors, withdrawing electron density from the lone pairs of O and N of the trioxy triazine. Back donation of electron density from Al metal to the organic layer is operative beneath the oxide layer. The dicyanate monomer doesn't adhere at all and desorbs quickly out of the interphase region on the substrate. It was concluded that thermosetting reaction of the prepolymer is thus hampered and the resulting network will be less dense near the substrate than in the bulk.

14.4
Miscellaneous Applications

Bisphenol-based aryl dicyanates containing phenyl phosphine oxide moieties and cyanate esters from various styrene-based polymers and copolymers are known to exhibit good flame retardant properties [39, 49, 86]. A large number of fire resistant cyanate ester compositions have been formulated [365–370]. The realistic brief period of multifunctional CEs is ascertained to be 2–5 min in most of the fire, smoke, and toxicity tests [371]. Such matrices are of obvious significance for application in aircraft interior structures. Polyfunctional CE-based formulations are used as heat resistant photoresists [372]. These photocurable coatings are resistant to moisture, temperature and high pressure [373]. Cyanate esters also find application in thin films [374–376], foams [161, 377], friction materials with improved antifade properties [352, 378, 379], and as binder for rare earth magnetic powders, which retards their oxidation [302, 380]. Polycyanurate-based wave guides are formed by spin coating of solutions of cyanate prepolymers followed by heat curing [381]. Advantages of such waveguides are high refractive index, low loss, and high T_g to lock permanently the polar orientations of an optically nonlinear dye. Fang invented an optically nonlinear polycyanurate by coupling diazo salt of dicyanovinylbenzene with bis(4-cyanatobenzyl) aniline [382]. Curing of the cyanate ester positions the NLO material after poling which retained the activity above 85 °C. The higher T_g of CE was conducive to reliable performance of the waveguides.

15
Outlook

The preceding review on the recent developments in the science and technology of cyanate esters clearly shows that this system emerges as a new generation of thermosetting polymer, encompassing several characteristics required of an ideal high performance matrix. The salient feature of addition curing through cyclotrimerization of cyanate groups without any volatile evolution is a highly desired for their use in void- free composites. A literature survey shows that cyanate esters have the added advantages of attractive physical, electrical, and mechanical properties. Thus, it is clear that CEs and PT resins possess significant advantages over existing high performance matrix resins for aerospace applications. It is well recognized that polycyanate esters show much lower out-gassing and exhibit higher heat tolerance than most epoxy resins. Several experimental and a few commercial CE resins are now available. Introduction of new cyanate ester resin has come to a state of stagnation. The present focus is on matrix modification and evolving new processing techniques for the existing systems. Practical problems in cyanate ester technology appears to be the difficulty in synthesizing scrupulously pure resins which offer reasonable shelf-life and predictable cure-profile, although a large number of cyanate systems with varying backbone structures and properties are designable. BACY has attracted a market due to its ease of synthesis, purification, and storability. However, most of the vital information pertaining to its synthesis, polymerization aspects, processing, and application are covered by patents and such information is to be generated at the user end. The attractive physical, mechanical, and processing characteristics of cyanate systems predict that the coming years will witness their emergence as the matrix of choice in many critical engineering areas, as this single system answers the majority of the serious problems faced by many of the current polymer matrices. Although the CE system is quite old, fundamental aspects of reaction mechanism, structure modification, blending, cure kinetics, reaction modeling, processing, etc., continue to evince interest of researchers. Development of CE-based technologies for a variety of applications in semiconductors, adhesives, foams, optical devices, etc., continues to be a subject of intense research and presently forms the subject of many patents.

A major hurdle to the widespread acceptance of CE systems as potential substitutes for the state-of-the-art materials for general purpose and specialty areas is their high cost. Technologists will naturally look forward to systems with costs comparable at least to the bismaleimides if not the epoxies. The worldwide production of CEs is less than 1000 tonnes. As Hamerton [166] has pointed out, the excessive cost of CEs arises from small batch production, adaptation of multipurpose equipment, and uneconomical recovery of insufficient amount of spent solvent. It is expected that this aspect is going to be given due consideration. Available information shows that the use of CE systems in aerospace application, an area where cost is only secondary to performance and weight-saving, is on the increase and they are phasing out the epoxies in structural components. It appears that, at many phases, these systems are in application trials. Since the CE system is quite young, information on such trial studies will be divulged only at a later time. In high speed circuitry, innovative research is likely to give CE systems a unique, dominant place in the near future. Another area that could soon catch up is high temperature resistant photoresist. The present studies on photocatalyzed curing of CEs are expected to pave the way for this. Application in nonlinear optics is yet another area with a high scope. One aspect that needs immediate redress is, perhaps, the relatively higher temperature needed for a perfect cure of cyanate esters. Although photocatalyzed curing appears promising, this can induce at best only the gelation of the system at ambient, the full cure invariably warranting heating to a higher temperature. Information on safety and health hazards, particularly of the thermally degraded products of CE systems, is yet to be made available.

Acknowledgments. The authors are grateful to Vikram Sarabhai Space Centre for permission to publish this article. They express their special thanks to Prof. I. Hamerton for sharing much information.

References

1. Lee H, Neville K (1967) Handbook of epoxy resins. McGraw Hill, New York
2. McAdams LV, Gannon JA (1986) Encyclopedia of polymer science and engineering. Wiley, New York, vol 6, p 322
3. Bosch A (1996) In: Salamone JC (ed) Epoxy resins, In: Polymeric materials encylcopedia. CRC press, Boca Rathon, vol 3, p 2246
4. Stover D (1994) High Perfom Comp, July/Aug:18
5. Arnold CA, Hergenrother PM, McGrath JE (1992) In: Lvigo T, Kinzig BJ (eds) Composite applications: the role of matrix, fiber and interface. VCH Publishers, chap 1
6. Bauer RS, Stewart SL, Stenzenberger HD (1993), Kirk-othmor encyclopedia of chemical technology, 4th edn. Wiley, New York, vol 7, p 33
7. Bauer RS, Filippove AG, Schlaudt LM, Breitigam WT (1987) SAMPE 32:1104
8. Takekoshi T (1990) Adv Polym Sci 94:1
9. Feger C, Khojasteh M, McGrath JE (1989) Polyimides: materials, chemistry and characterisation. Elsevier
10. Wilson D (1990) Brit Polym J 20:405
11. Rogers ME, Moy TM, Kim YJ, McGrath JE (1992) Mat Res Soc Symp Proc 264:13
12. Volksen W (1990) Recent advances in polyimides and other high performance polymers. Workshop sponsored by American Chemical Society, Polymer Division
13. De Abajo J, de la Campa JG (1998) Progress in polyimide chemistry. Adv Poly Sci 140:23–59
14. Sat M (1997) Polyimides. In: Olabisi O (ed) Handbook of thermoplastics. Marcel Dekker, New York
15. Mittal KL (ed) (1985) Polyimides, synthesis, characterisation and applications. Plenum, New York
16. Lin SC, Pearce EM (1994) High performance thermosets: chemistry, property and applications. Hanser, Munich
17. Chandra R, Rajabi L (1997) J Macromol Sci, Rev Macromol Chem Phys C37:61
18. Mison P, Sillion B (1998) Progress in polyimide chemistry 1. Adv Polym Sci 140:137
19. Stenzenberger HD (1994) Addition polyimides. Adv Polym Sci 117:163
20. Stenzenberger HD, Romer W, Hergenrother PM, Jensem B, Breitigam W (1990) SAMPE J 26:75
21. San Diego CA (March 1994) Proc Polym Mater Sci Eng, vol 70
22. Pilato LA, Michno J (1994) Advanced composite materials. Springer, Berlin Heidelberg New York
23. Pilato LA, Michael JM (1994) Advanced composite materials, Springer, Berlin Heidelberg New York, Chap 2

24. Gorham W (1966) J Polym Sci, PtA-1, 4:3027
25. Katsman HA, Mallon JJ, Barry WT, (1995) J Mat Mater 21
26. Cava MP, Mitchell MJ (1967) Cyclobutadiene and related compounds. Academic Press, NY, Chap 6
27. Serafini TT, Delvigs P, Lightsey G (1972) J Appl Polym Sci 16:905
28. Kopf PW, Little AD (1988) Phenolic resins In: Mark HF, Bikales NM, Overberger CG, Menges G (eds) Encyclopedia of polymer science and engineering, 2nd edn. Wiley, vol 11, p 45
29. Knop A, Pilato LA (1985) Phenolic resins: chemistry, applications and performance, future directions. Springer, Berlin Heidelberg New York
30. Fukuda A (1996) Phenolic resins. In: Salamone JC (ed) Polymeric materials encyclopaedia. CRC Press, Florida, vol 7, p 5035
31. Nair CPR, Bindu RL, Ninan KN (1997) Recent advances in phenolic resins, In: Metals, materials and processes. Meshap Science Publishers, Mumbai, India, vol 9(2), p 179
32. Knop A, Schieb W (1979) Chemistry and applications of phenolic resins. Springer, Berlin Heidelberg New York
33. Bindu RL (1999) Modified Phenolic Resins and Composites, Ph. D. thesis, Kerale University, Trivandrum, India
34. Grigat E, Putter R (1967) Angew Chem Int Ed 6:206
35. Falchetto A (1998) PCT Int Appl WO 98 23,584
36. Grigat E, Schminke HD, Putter R (1965) Ger 1,190,184
37. David RWH Jr, Das S (1996) PCT Int Appl WO 96 23,013
38. George GD, William BB, Raymond SJ (1993) US 5,264,500
39. Abed J, Mercier R, McGrath JE (1997) J Polym Sci Polym Chem 35(6):977
40. Watabe H, Shibata M (1996) Jpn Kokai Tokkyo Koho JP 08 92,192
41. Watabe H, Hayashi T, Shibata M (1996) Eur Pat Appl EP 739,879
42. Bao J, Tang B (1998) Reguxing Shuzhi 13(1):18
43. Martin D (1964) AngewChem 76:303
44. Grigat E, Putter R (1964) Ber 97(11):3012
45. Hamerton I (ed) (1994) Chemistry and technology of cyanate ester resins. Blackie Academic and Professional, Glasgow, UK
46. Abed JC, Mercier R, Srinivasan SA, McGrath JE (1992) Polym Prepr Amer Chem Soc Div Polym Chem 33(2):233
47. Shimp DA, Vanderup JT (1991) Eur Pat. Appl E 449,593
48. Nair CPR, Bindu RL, Joseph VC (1995) J Polym Sci Polym Chem 33:621
49. Kern W, Cifrain M, Schroder R, Hummel K, Mayer C, Hofstotter M (1998) Eur Polym J 34(7):987
50. Stroh R, Gerber H (1960) Angew Chem 72:1000
51. Millard TG, Puckett PM (1994) Polym. Mater Sci Eng 71:625
52. Chaplin A, Hamerton I, Howlin BJ, Barton JM (1994), Macromolecules 27(18):4927
53. Smith CD, Webster HF, Wightman JP, McGrath JE (1991) High Perform Polym 4(3):211
54. Lyon RE (1996) SAMPE J 32(3):29
55. Lin B, Kinne PM (1998) Int SAMPE Symp Exhib Book 2, vol 43, p 1560
56. Lin B, Kinne PM (1998) SAMPE J 43(2):1560
57. Walters R, Lyon R (1997) SAMPE J 42(2):1335
58. Lin B (1999) Proceedings 44th International SAMPE Symposium, May 23–27, p 1422
59. Giannelis E (1996) Polymer layered silcate nanocomposites. Adv Mater 8:29
60. Gilmann JW, Harris R Jr, Hunter D (1999) Proceedings 44th International SAMPE Symposium, May 23–27, p 1408
61. Lyon RE, Waters RN, Gandhi S (1998) Int SAMPE Symp Exhib Book 2, Vol 43, p 1586
62. Dona Mathew, Nair CPR, Ninan KN (1997) Proc 9th Kerala Sci Cong, Damodaran AD (ed), Sci Tech Dept, Trivandrum, India, p 286
63. Dona Mathew, Nair CPR, Ninan KN (2000) Polym Int 49:1
64. Fyfe CA, Niu J, Mok K (1995) J Polym Sci 33:1191

65. Rogers JK (1988) Plast Technol Aug:31
66. Rogers JK, Gabriele MC (1989) Plast Technol Jul:25
67. Marcos-Fernandez A, Posadan P, Rodriguez A, Gonzalez L (1999) J Polym Sci Polym Chem Ed 37:3155
68. Das S, Prevorsek DC, De Bona BT (1989) 21st Int SAMPE Tech Conf, September 1989
69. Das S, Parsippany, Prevorsek DC (1989) US Pat 4,831,086
70. Still JK, Nelb RG, Norris SO (1976) Macromolecules 9:3
71. Marcos FA, Posados P, Gonzalez L, Rodriguez A (1998) Polym Prepr ACS Div Polym Chem 39(2):574
72. Gandhi S (1999) Proc Int Conf Fire Saf 27:375
73. Snow AW, Buckley LJ (1997) Macromolecules 30:394
74. Snow AW, Buckley LJ, Armistead JP (1998) Polym Prepr Am Chem Soc Div Polym Chem 39(2):788
75. Conley RT (1970) Thermal stability of polymers. Marcel Decker, chap 11
76. Cozzens RF, Walter P (1987) J Appl Polym Sci 34:601
77. Miks,MW, Shigly JK (1997) US Pat 5,645,219
78. Das S, Prevorsek DC, DeBona BT (1990) Modern Plast Int, June p.64
79. Delano CB, McLeod AH (1979) NASA-CR-159724, FR-79-25/AS
80. Allen P, Childs W (1993) 38th Int SAMPE Symp, p 533
81. Mathew D, Nair CPR, Ninan KN (1998) In: Srinivasan KSV (ed), Proceedings of Macro-98, IUPAC International Conference, Macromolecules New Frontiers, Allied Publishers, New Delhi, vol 2 p 948
82. Mathew D, Nair CPR, Ninan KN (1999) Eur Polym J 36:1195
83. Mathew D, Nair CPR, Ninan KN (2001) Eur Polym J (in press)
84. Bindu RL, Nair CPR, Ninan KN (2000) J Polym Sci Polym Chem 38:641
85. Ikeguchi N, Nozaki M (1999) Japan Kokai Tokkyo Koho JP 11 124,433; Chem Abs 130:338,839
86. Gilman JW, Haris RH Jr, Brown JE (1997) 42nd Int SAMPE Symp, May 4–8, p 1052
87. Gupta AM, Macosko CW (1993) Macromolecules 26(10):2455
88. Shimp DA (1986) Proc Am Chem Soc Div Polym Mater Sci Eng 54:107
89. Fang T, Shimp DA (1995) Progr Polym Sci 20:61
90. Barton JM, Greenfield DC, Hamerton I, Jones JR (1991) Polym Bull 25:475
91. Fyfe CA, Niu J, Rettig SJ, Burlinson NE, Reidsema CM, Wang DW, Poliks M (1992) Macromolecules 25:6289
92. Fang T, Houlihan FM (1994) Polym Prepr 35:535
93. Simon SL, Gillham JK (1993) J Appl Polym Sci 47:461
94. Grenier-Loustalot MF, Lartigau C, Grenier P (1995) Eur Polym J 31:1139
95. Grenier-Loustalot MF, Lartigau C, Metras F, Grenier P (1996) J Polym Sci Polym Chem 34:2955
96. Grenier-Loustalot MF, Lartigau C (1997) J Polym Sci Polym Chem 35:3101
97. Cunningham ID, Brownhill A, Hamerton I, Howlin BJ (1994) J Chem Soc Perkin Trans 2:1937
98. Deng Y, Martin GC (1996) Polymer 37:3593
99. Deng Y, Martin GC (1997) J Appl Polym Sci 64:115
100. Bauer M, Bauer J (1994) In: Hamerton I (ed) Chemistry and technology of cyanate ester resins. Blackie Academic and Professional, Glasgow, UK, p 58
101. Balko JW, Gotro JT (1992) Polym Mater Sci Eng 66:449
102. Galy J, Gerard JF, Pascault JP (1991) In: Abadie MJM, Sillion B (eds) Polyimides and other high temperature polymers, Elsevier Science, Amsterdam, p 245
103. Ustyak VV, Manzenkov AV, Lukin PM (1996) Zh Obshch Ksim (Russian) 66(11):1860
104. Cooper JB, Vess TM, Campbell LA, Jensen BJ (1996) J Appl Polym Sci 62:135
105. Shimp DA, Craig WM Jr (1989) 34th Int SAMPE Symp, May 8–1, p 1336
106. Shimp DA (1987) 32nd Int SAMPE Symp April 6–9, p 1063
107. Bauer M (1980) Thesis, Academy of Sciences of the GDR, Berlin

108. Liu H, George GA (1996) Polymer 37(16):3675
109. Owusu OA, Martin GC, Grotto JT (1991) Polym Eng Sci 31(22):1604; (1992) Polym Eng Sci 32:535
110. Owusu OA, Martin GC, Gottro JT (1991) ACS Polym Mater Sci Eng Prepr 65:304
111. Nair CPR, Mathew D, Ninan KN (1996) In: Salamone JC (ed) Cyanate ester resins. Polymeric materials encyclopedia, vol 2. CRC Press, Florida, p 1625
112. Bauer M, Bauer J (1989) Makromol Chem Macromol Symp 30:1
113. Gupta AM, Macosko CW(1991) Macromol Chem Macromol Symp 45:105
114. Korshak VV, Pankratov VA, Ladovskaya AA, Vinogradoa SV (1978) J Polym Sci Chem 16:1697
115. Nair CPR, Mathew D, Ninan KN (1999) J Polym Sci Chem 37:1103
116. Bauer M, Bauer J, Kuhn G (1986) Acta Polym 37:218,221, 715
117. Owusu OA, Martin GC, Gotro JT (1996) Polymer 37(21):4869
118. Graeme GA, Gregory CA, Liu H, Vassallo T (1996) Polym Mater Sci Eng,74:90
119. Niu J (1994) Diss Abstr Int B 55(2):396
120. Simon SL (1992) PhD Thesis, Princeston University
121. Alla C (1992) Thesis, Université Pierre et Marie Curie, Paris
122. Georjon O, Galy J, Pascault JP (1993) J Appl Polym Sci 49:1441
123. Mirco V, Mechin F, Pascault JP (1994) Polym Mater Sci Eng 71:688
124. Shimp DA, Ising SJ (1992) ACS PMSE Preprints 66:504
125 . Alistair J (1998) Polym Compos Rapra Technol Ltd 6(3):121
126. Grant JT, Dunwoody N, Jakubek V, Lees AJ (1998) Polym Prepr ACS Div Polym Chem, 39:687
127. Grant GT, Dunwoody N, Jakubek V, Lees AJ (1998) Polym Compos Rapra Technol Ltd 6(1):47
128. McComick FB (1994) Polym Mat Sci Eng 71:680
129. McComick FB (1992) Polym Mat Sci Eng 66:460
130. Kotch TG, Lees AJ, Fuerniss SJ, Papathomas KI (1995) Chem Mater 7:801
131. Kotch TG, Alistair LJ, Fuerniss S J, Papathomas KI (1992) Polym Mat Sci Eng 66:462
131a. Coats AW, Redfern JP (1964) Nature, 201:68
132. Hamerton I, Takeda S (2000) Polymer 41:1647
133. Snow AW (1994) In: Hamerton I (ed) Chemistry and technology of cyanate ester resins. Blackie Academic and Professional: Glasgow, UK, p 61
134. Kenny JM (1994) J Appl Polym Sci 51(4):761
135. DiBenedetto AT (1987) J Polym Sci Phys 25:1949
136. Pascault JP, Williams RJJ (1990) J Polym Sci Phys 28:85
137. Hale A, Macosko CW, Bair HE (1991) Macromolecules 24:2610
138. Bauer M, Gnauck R (1987) Acta Polym 38:658
139. Venditti RA, Gillham JK (1997) J Appl Polym Sci 64:3
140. Chen YT, Macosko CW (1996) J Appl Polym Sci 62 :567
141. Bauer J, Hoeper L, Bauer M (1998) Macromol Chem Phys 199 (11):2417
142. Zukas WX, James VT (1995) 53rd Ann Tech Conf Soc Plast Eng 2:2761
143. Lu MC, Hong JL (1994) Polymer 35:2822
143a. Havlicek I, Dusek K (1987) In: Sedlacek B, Kahovec J (eds) Crosslinked epoxies. Walter de Gruyter, Berlin, p 417
143b. Lin RH, Su AC, Hong JL (1999) J Appl Polym Sci 73:1927
144. Rabinowitch E (1937) Trans Faraday Soc 33:1245
145. Gupta AM (1991) Macromolecules 24:3459
146. Mathew D, Nair CPR, Ninan KN (2000) J Appl Polym Sci 77:75
147. KhannaYP, Kumar R, Das S (1989) Polym Eng Sci 29(20):1488
148. Hay JN (1994) In: Hamerton I (ed) Chemistry and technology of cyanate esters. Blackie Academic and Professional, Glasgow, chap 6
149. Ising SJ, Shimp DA, Christenson JR (1989) 3rd Int SAMPE Electronics Conf, 20–22 June, pp 360–370

150. Itoya K, Kakimoto M, Imai Y (1994) Polymer 35:6
151. Vaughan JG, Lackey E, Theobold D, Blossom M (1995) 40th Int SAMPE Symp May 8–1, p 1293
152. Georjon O, Galy J (1997) J Appl Polym Sci 65:2471
153. Boyle M, Lee F (1989) 21st Int SAMPE Tech Conf 21:294
154. Ciba Geigy RTM Bulletin on AroCY cyanate esters, Ciba-Geigy Speciality Resin Division, Kentucky 40269, USA
155. Shimp DA, Christensen JR (1990) In: Hornfield HL (ed) Plastics, metals and ceramics SAMPE, pp 81–94
156. McConnell VP (1992) Advanced Composites May/Jun
157. Cinquin J, Abjean P (1993) 38th Int SAMPE Symp Exhib, p 1539–551
158. Couch BP, McAllister LE (April 1990) 35th Int SAMPE Symp, p 2298
159. Nair CPR, unpublished results
160. Speak SC, Sitt H, Fuse RH (1991) 36th Int SAMPE Symp, p 336
161. Wang Y-S, Kuo C-C (1991) Int SAMPE Symp Exhib 36(2):1430
162. Mackenzie PD, Malhotra V (1994) In: Hamerton I (ed) Chemistry and technology of cyanate esters. Blackie Academie and Professional, Glasgow, chap 9
163. Zeng S, Hoisington M, Seferis JC, Shimp DA (1992) 37th Int SAMPE Symp, p 348
164. Shimp DA, Ising SJ (1992) ACS PMSE Prepr 66:504
165. Walsh AT (1991) Metals Materials 606
166. Hamerton I (1998) High Perform Polym 10:163
167. Joardar SS, Srinivasan SA, Priddy DB, Mc Grath JE, Ward TC (1994) ACS Polym Prepr 208:332
168. Hillermeier RW, Hayes BS, Seferis JC (1999) Polym Compos 20 (1):155
169. Li C, Kohn J (1986) Biomaterials 7:176
170. Shimp DA, Wentworth JE (1992) 37th Int SAMPE Symp March 9–2, p 293
171. Bauer M, Bauer J, Ruhman R, Kuhn G (1989)Acta Polym 40:397
172. Bauer J, Bauer M (1990) J Macromol Sci Chem A27:97
173. Meyer G, Bauer J, Bauer M (1994) Polym Mater Sci Eng 71:797
174. Pascault JP, Galy J, Mechin F (1994) In: Hamerton I (ed) Chemistry and technology of cyanate esters. Blackie Academic and Professional, Glasgow, chap 5
175. Bauer M, Tanzer W, Much H, Ruhman R (1989) Acta Polym 40:335
176. Bauer J, Bauer M (1990) J Polym Sci Polym Chem Ed 28:97
177. Bauer J, Bauer M (1990) Acta Polym 41:535
178. Fyfe CA, Niu J, Rettig SJ, Wang DW, Polik MD (1994) J Polym Sci Polym Chem 32:2203
179. Martin MD, Ormaetxea M, Harismendy I, Remiro PM, Mondragon I (1999) Eur Polym J 35(1):57
180. Fainleib AM, Shantali T, Sergeeva L (1995) Plastmassy1:16
181. Fainleib AM, Seminovych HM, Slinchenko EA, Brovko AA, Sergeeva LM (1998) In: Srinivasan KSV (ed) Proceedings of Macro-98, IUPAC International Conference, Macromolecules New Frontiers, Allied Publishers, New Delhi, p 818
182. Kim BS (1997) J Appl Polym Sci 65:85
183. Wang CS, Lee MC (1999) J Appl Polym Sci 73(9):1611
184. Penczek P, Kaminska W (1990) Adv Polym Sci 97:41
185. Sakamoto K, Higuchi Tkasai Y (to Matsushita Electrical works) (1976) Jpn Pat Appl 76,39,770; Chem Abs 85:47,799
186. Mitsubishi Gas Chemical Co (1985) Jpn Pat Appl 85,125,661; Chem Abs 103:197,100
187. Mitsubishi Gas Chemical Co (1981) Jpn Pat Appl 81,141,310; Chem Abs 96:53,279
188. Urabe H, Ikeguchi N (1998) Japan Kokai Tokkyo Koho JP 10,315,385
189. Urabe H, Ikeguchi N (1999) Japan Kokai Tokkyo Koho JP 11 54,886
190. Urabe H, Ikeguchi N, Kato Y (1998) Japan Kokai Tokkyo Koho JP 10 337 807 191; Urabe H, Ikeguchi N, Kato Y, Japan Kokai Tokkyo Koho JP 10 337 809 192
192. Eurich J (1991) Proc Int Electron Packaging Conf (IEPS), San Diego, vol 2, p 1094

193. Lee FW, Baron KS, Boyle MA (1991) US Pat 5,045,609
194. Shimp DA (1994) In: Hamerton I (ed) Chemistry and technology of cyanate ester resins. Blackie Academic and Professional, Glasgow, UK, chap 10, p 284
195. Chau M, Burkhart M, Donald A (1999) US Pat 5,855,821
196. Konarski M, Szczepniak Z (1999) PCT Int Appl WO 99 05 196
197. Blaire C, Zakrzewski J (1992) Design of optical instruments. SPIE 1690:300
198. Wada M, Takeda T(1999) Japan Kokai Tokkyo Koho JP 11 106,481; Chem Abs 130:312,956
199. Wada M, Taketa T (1999) Japan Kokai Tokkyo Koho JP 11 106,480; Chem Abs 130:312,955
200. Mizuno Y, Sase S, Takano M (1999) Japan Kokai Tokkyo Koho JP 11 106,613
201. Nguyen GP, Edwards C (1999) US Pat 5,912,316
202. Ikeguchi N, Nozaki M (1999) Japan Kokai Tokkyo Koho JP 11 124,426; Chem Abstr: 130:353,423
203. Ikeguchi N, Nozaki M (1999) Japan Kokai Tokkyo Koho JP 11 124,434, Chem Abs 130:338,840
204. Hayashi T, Nakajima N (1999) Japan Kokai Tokkyo Koho JP 11 140,275
205. Hayashi T, Nakajima N (1999) Japan Kokai Tokkyo Koho JP 11 140,276
206. Takeda T, Matsuda Y, Murayama R, Okubo M (1998) Japan Kokai Tokkyo Koho JP 10 130,465 ; Chem Abs 129:55,252
207. Takeda T, Matsuda Y, Murayama R, Okubo M, Okubo H (1998) Japan Kokai Tokkyo Koho JP10 139,858 ; Chem Abs 129:55,262
208. Taketa T, Matsuda Y, Murayama R, Okubo H (1998) Japan Kokai Tokkyo Koho JP 10 163,232
209. Krumweide GC, Akau RL (1996) AIP Conf Proc, Part 2, 361:883
210. Mathew D, Nair CPR, Ninan KN (1999) J Appl Polym Sci 74:1675
211. Jianwen B, Xiangbao C, Xiao-Su Y (1999) 44th Int SAMPE Symp, p 1134
212. Lin SC, Pearce EM (1993) In: High performance thermoset, chemistry, properties, applications. Hanser, Munich
213. Itoh M (1994) Purasuchikkusu (Jp) 45(9):38; Chem. Abs (1994) 121:206,985
214. Hamerton I (1996) High Perform Polym 8:83
215. Gaku M (1978)US Pat 4,110,364
216. Ayano S (1985) Kunstoffe 75:475
217. Nair CPR, Francis T (1999) J Appl Polym Sci 74:2737
218. Enoki T, Takeda T, Ishii K (1995) Netsu Kokasei Jushi 16(1):1
219. Stenzenberger HD (1991) In: Wilson D, Stenzenberger HD, Hergenrother P (eds) Polyimides. Blackie, Glasgow, UK
220. Barton JM, Hamerton I, Jones JR (1992) Polym Int 29:145
221. Hong JL, Wang CK, Lin RH (1994) J Appl Polym Sci 53:105
222. Sperling LH (1994) In: Klempner D, Sperling LH, Ultracki LA (eds) Interpenetrating polymer networks. Am Chem Soc
223. Hamerton I, Jones JR, Barton JM (1994) PMSE Polymer Prepr 71:807
224. Reyx D, Campistron I, Caillaud C, Villate M, Cavedon A (1995) Makromol Chem 196:775; Cunningham ID, Brownhill A, Hamerton I, Howlin BJ (1997) Tetrahedron 53:13,473
225. Chaplin A, Hamerton I, Howlin BJ, Barton JM (1994) Macromolecules 27:4927
226. Barton JM, Hamerton I, Jones JR, Stedman JC (1996) Polymer 37(20):4519
227. Barton JM, Hamerton I, Jones JR (1993) Polym Int 31:95
228. Barton JM, Chaplin A, Hamerton I, Howlin BJ (1999) Polymer 40:5421
229. Barton JM, Hamerton I, Chaplin A (1998) PCT Int Appl WW 98 18,755
230. Hatao T, Yamaji T, Ohtori T (1999) Japan Kokai Tokkyo Koho JP 11 116,671
231. Hefner RE Jr, Jackson L (1987) US Pat 4,683,273
232. Hefner RE Jr (1989) Braz Pedido PI Br 8703,011
233. Hefner RE Jr (1987) US Pat 4,680,378

234. Hwang HJ, Wang CS (1998) J Appl Polym Sci 68:1199
235. Nair CPR, Francis T, Krishnan K, Vijayan TM (1999) J Appl Polym Sci 74:2737
236. Chaplin A, Hamerton I, Herman H, Mudhar AK, Shaw SJ (2000) Polymer 41:3945
237. Bauer M, Bauer J, Jahrig S (1991) Makromol Chem Macromol Symp 45:97
238. Kohn J, Langer R (1989) Macromolecules 22:2029
239. Pankratov VA, Maiorova AA, Korshak VV, Vinogradova SV (1975) Vysokmol Soed A17:2189
240. Arnold C, Mackenzie P, Malhotra V, Pearson D, Chow N, Hearn M, Robinson G (1992) 37th International SAMPE Symp March 9–2, p 128
241. Cao ZQ, Mechin F, Pascault JP (1994) Polym Int 34:41
242. Cao ZQ, Mechin F, Pascault JP (1994) Polym Mater Sci Eng Prepr 70:91
243. Uhlig C, Bauer J, Bauer M (1995) Macromol Symp 93:69
244. Srinivasan SA, McGrath JE (1993) High Perform Polym 5:259
245. Srinivasan SA, Joardar SS, Priddy DB Jr, Ward TC, Mcgrath JE (1994) Proc 39th Int SAMPE Symp and exhib, Anaheim, CA, p 60
246. Srinivasan SA, Mcgrath JE (1997) J Appl Polym Sci 64:167
247. Hourston DJ, Lane JM, (1992) Polymer 33:1379
248. Almen G, Mckenzie P, Malhotra V, Maskell R (1989) 21st Int SAMPE Tech Conf, Sept, p 304
249. Yang PC, Pickleman DM, Woo EP (1990) 35th Int.SAMPE Symp, p 1131
250. Borrajo J, Riccardi CC, Williams RJJ, Cao ZQ, Pascault JP (1995) Polymer 36:3541
251. Cao ZQ, Mechin F, Pascault JP (1996) In: Riew CK (ed) Toughened plastics II – novel approaches in science and engineering. Am Chem Soc, Advances in Chemistry Series, Washington, 252:177
252. Hillermeier RW, Hayes BS, Seferis JC (1998) In: Williams KR (ed) Proc 26th Conf North Am Therm Anal Soc, Omni Press: Madison, Wis, p 99
253. Srinivasan SA, McGrath JE (1998) Polymer 39(12):2415
254. Porter DS, William CA, Srinivasan SA, McGrath JE, Ward TC (1995) Proc 18th Annual Meet Adh Soc, p 102
255. Rau AV (1997) Diss Abstr Int B 57(10):6354
256. Srinivasan SA, Lyle GG, McGrath JE (1993) Int SAMPE Symp Exhib 38:28
257. Ardakani AA, Gotro JT, Hedrick NM, Shaw JM, Viehbeck A (1994) Eur Pat Appl EP 581,314
258. Porter DS, Brown JM, Srinivasan SA, McGrath JE, Ward TC (1994) Polym Mater Sci Eng 71:817
259. Diberardino M, Pearson R (1993) Int SAMPE Tech Conf 25:502
260. Lee BK, Kim SC (1995) Polym Adv Technol 6(6):402
261. Pollack S, Fu Z (1998) Polym Prepr Am Chem Soc Div Polym Chem 39(1):452
262. Anon (1993) Res Disc (UK) 345:69
263. Srinivasan SA, Joardar SS, Priddy DB Jr, Ward TC, McGrath JE (1993) Polym Mater Sci Eng 70:93
264. Srinivasan SA, McGrath JE (1993) Proc 16th Annu Meet Int Symp Interphase, Adh Soc, p 407
265. Srinivasan SA, McGrath JE (1993) SAMPE Q 24(3):25
266. Hwang JW, Park SD, Cho K, Kim JK, Park CE, Oh TS (1997) Polymer 38(8):1835
267. Hwang JW, Cho K, Park CE, Huh W (1999) J Appl Polym Sci 74:33
268. Hamerton I, Heald CR, Howlin BJ (1996) J Mater Chem 6 :311
268a. Hamerton I, Hay HN (1998) Polym Int 47:465
269. Chang JY, Hong JL (1998) Polymer 39(26):7119
270. Brown JM, Srinivasan SA, Rau AV, Ward TC, McGrath JE, Loos AC, Hood DK (1996) Polymer 37(9):1691
271. Srinivasan SA, Joardar SS, Kranbeuhl D, Ward TC, McGrath JE (1997) J Appl Polym Sci 64(1):179
272. Rau AV, Srinivasan SA, McGrath JE, LoosAC (1998) Polym Compos 19(2):166

273. Parker DG, Wheatley GW (1994) Polym lnt 3(33):321
274. McGrail PT, Jenkins SD (1993) Polymer 34(4):677
275. Srinivasan SA, Rau AV, Loos AC, McGrath JE (1994) Polym Mater Sci Eng 71:750
276. Jenkins SD (1992) Proc Eur Conf Adv Mater, 2nd edn, vol 3, p 268
277. Hirohata T, Kuroda M, Besso H, Nishimura A, Inoue T (1998) SEI Tekunikaru Rebyu 153:128
278. Hirohata T, Kuroda M, Nishimura A, Inoue T, Nippon F (1998) Zairyo Gakkaishi 24(2):69
279. Kiefer J, Hilborn JG, Hedrick JI, Cha HJ, Yoon DY, Hedrick JC (1996) Macromolecules 29:8546
280. Kiefer J, Hedrick JL, Hilborn JG (1999) Adv Polym Sci 47:161
281. Penczek P, Mirkowska B (1993) Ang Makromol Chem 205:19
282. Kim YS, Sung C (1999) Macromolecules 32:2334
283. Mormann W, Zimmermann J (1995) Macromol Symp (Polymer Networks) 93:97
284. Mormann W, Zimmermann JG (1996) Macromolecules 29:1105
285. Mormann W, Zimmermann JG (1995) Makromol Symp 93:97
286. Mormann W, Zimmermann JG (1995) Liq Cryst 19(2):227
287. Mormann W, Zimmermann JG (1995) PCt Int Appt WO 9507, 268
288. Mormann W (1995) Trend Poly Sci 3:255
289. Clingman SR, Ober CK (1995) Polym Mater Sci Eng 72:238
290. Mahlstedt S, Bauer M (1994) Polym Mater Sci Eng 71:801
291. Ou J, Hong YL, Yen FS, Hong JL (1995) J Polym Sci Polym Chem 33(2):313
292. Wang YH, Hong YH, Yang FS, Hong JL (1994) Proc Am Chem Soc Div Polym Mater Sci Eng Preprints 71:678
293. Holland WR, Fang T (1992) PMSE Preprints 66:500
294. Ober CK, Korne H, Shiota A, Laus M (1996) Angew Makromol Chem 240:59
295. Gaku M (1994) ACS Polym Div Polym Mater Sci Eng 71:621
296. Kasehagen LJ, Haury I, Macosko CW, Shimp DA (1997) J Appl Polym Sci 64:107
297. Kanoute P, Dumontete H, Hauviller C, Nicquevert B (1998) In: Gibson G (ed) Proc 7th Int Conf Fibre Reinf Compos, Cambridge, UK, p 317
298. Zacharia RE, Simon SL (1997) J Appl Polym Sci 64:127
299. Chung K, Seferis JC (1998) Int SAMPE Symp Exhib 43, Book1, p 387
300. Lee BL, Holl MW (1994) Proc 9th Am Soc Compos Tech Conf, p 1045
301. Hnatowicz VV, Vacik J, Svorcik V, Rybka V, Fink D, Klett R (1997) Czech J Phys 47(2):255
302. Shimp DA, Southcott M (1993) 38th Int SAMPE Symp Exhib, p 370
303. Gutman EM, Grinberg A, Ribak EA, Petronius I (1997) Polym Compos 18(4):561
304. Donnell T, Lewis D, Clark M, Dolgin B, Sokolowski W (1993) 38th Int SAMPE Symp Exhib, p 1566
305. Hahn DR, Joardar SS, Srinivasan SA, Ward TC (1993) Polym Mater Sci Eng 69:74
306. Parvatareddy H, Tsang PHW, Dillard DA (1996) Compos Sci Technol 56(10):1129
307. Kasehagen LJ, Haury I, Macosko CW, Shimp DA (1996) J Appl Polym Sci (1996) 64(1):107
308. Wang JZ, Parvatareddy H, Chang T, Iyengar N, Dillard DA, Reifsnider KL (1995) Compos Sci Technol 54(4):405
309. Gendre T, Esnault R, Guedra DD (1996) Eur SPAgency (Spec Publ) ESA Sp-386, Proc Int Conf on Spacecraft Structures, Materials and Mechanical Testing, vol 2, p 507
310. Arnold FE, Guy A (1999) Proceedings 44th International SAMPE Symp, May 23–27, p 1575
311. Korshak VV (1974) Vysokmol Soed A16:15
312. Korshak VV (1975) Iza Akad Nauk Gruz SSR, ser Khim 2:451
313. Shimp DA, Ising SJ (1991) Proc ACS Div Polym Mater Sci Eng 66:504
314. Shimp DA, Chin B (1994) In: Hamerton I (ed) Chemistry and technology of cyanate ester resins. Blackie Academic and Professional, Glasgow, UK, chap 8, p 230
315. High Performance Plastics (July 1994) p 10

316. Deutsch A (1996) IEEE Trans Comp Packag Manuf Technol B19:331
317. Sase S, Mizuno Y, Sugimura T, Negishi H (1999) Japan Kokai Tokkyo Koho JP 11 124,451
318. Shimp DA (1994) ACS PMSE Prepr 71:561
319. Husson FD Jr, Neff B (1998) PCT Int Appl WO 98 10,920
320. Hayashi T (1997) Jpn Kokai Tokkyo Koho JP 09 278,882
321. Snow AW, Buckley LJ (1996) US Pat Appl 621,149
322. Minamimura K (1997) Jpn Kokai Tokkyo Koho JP 09 104,753
323. Wakamatsu H, Suemasa H (1997) US Pat 5,612,512
324. Papathomas KI (1994) Eur Pat Appl EP 604,823
325. Nakada T, Tsunemi H (1994) Jpn Kokai Tokkyo Koho JP 0649, 205
326. Wang PC, Kel S, Donald R (1994) US Pat 5,340,912
327. Papathomas KI (1994) US Pat 5,292,861
328. Motoori S (1998) Japan Kokai Tokkyo Koho JP 10 287,809
329. Kulesza JD, Markovich VR, Papathomas KI, Sabia JG (1999) US Pat 5,887,345
330. Sase S, Mizuno Y, Sugimura T, Negishi H (1999) Japan Kokai Tokkyo Koho JP 11 21,506
331. Sase S (1999) Japan Kokai Tokkyo Koho JP 11 21,507
332. Mizuno Y, Sase S, Kuritani H (1998) Japan Kokai Tokkyo Koho JP 10 273,532
333. Mizuno Y, Sase S, Kuritani H (1998) Japan Kokai Tokkyo Koho JP 10 273,533
334. Grayson MA, Fry CG (1994) Proc SEM Spring Conf Exp Mech Soc for Expt Mech, p 455
335. Galloway DP, Grosse M, Ngugen MY, Burkhart N (1995) IEEE/CPMT 17th Int Electron Manuf Technol Symp, p 141
336. Knouff B, Tompkins SS, Jayaraman N (1993) 25th Int SAMPE Tech Conf, p 610
337. Bryte Techol Inc (1993) 408/946-6477-No.21, SAMPE J, Sept/Oct 29(5)
338. Harvey JA (1993) In: Updadhaya Kamleswar (ed) Process Fabr Appl Adv Compos Proc Conf, ASM, Materials Park, Ohio, p 199
339. Robitaille S, Patz G, Johnson S (1994) Proc Int Workshop on Adv Mater for High Precision Detectors, p 115
340. Ikeda C, Ozaki T, Tsuneda S, Uchukozo (1996) Zairyo Shinpojumu (Jp) 12:134
341. Epstein G, Ruth S (1996) SAMPE J 32:1
342. Smaldone PL (1995) Int SAMPE Tech Conf 27:387
343. Hoffman CN, Snyder BA (1996) 41st Int SAMPE Symp, p 814
344. Almen GR, Mackenzie PD, Malhotra V, Maskel RK (1991) 23rd Int SAMPE Tech Conf, p 947
345. Willis PB, Coulter DR (1993) Proc SPIE Int Soc Opt Eng, p 127
346. Reed RP, Walsh RP (1994) Adv Cryogen Eng Part B 40:1129
347. Snow AW, Armistead JP (1994) In: Liechti KM (ed) Proc 17th Annu Meet Symp Part Adhes, Adh Soc Texas, p 53
348. Snow AW, Armistead JP (1995) J Adhes 52(1/4):223
349. Mehrkam P, Cochran R (1994) 26th Int SAMPE Tech Conf, p 215
350. Shimp DA, Ising SJ (1990) 34th Int SAMPE Symp, p 1045
351. Headifen RN, Cupta S, Okey D (1994) Compos Sci Technol 51(4):531
352. Ooya K, Sayama N, Kamioka N, Shibuya S, Nagakawa T, Yamashita M (1995) Jpn Kokai Tokkyo Koho JP 07,277,844
353. Iwashita T, Sawada Y (1996) Jpn Kokai Tokkyo Koho JP 08, 156,110
354. Nguyen N, Le KC (1995) US Pat Appl 430,427
355. Shimp DA (1991) Adhesion Soc Meet, Clearwater, Feb, p 16
356. Possart W, Dieckhoff S, Fanter D, Gesang T, Hartwig A, Hoper R, Schlett V, Hennemann OD (1995) Proc 18th Annu Meet Adhes Soc, p 15
357. Takeda H, Shiobara T (1997) Jpn Kokai Tokkyo Koho JP 09 100,349
358. Nguyen N (1993) PCT Int Appl WO 93.17,059
359. Nguyen N (1992)US Pat 5,155,066
360. Okada Y, Nojiri H (1994) Jpn Kokai Tokkyo Koho JP 06 256,743
361. Akahori R, Nagano K (1998) Japan Kaokai Tokkyo koho JP10 147,643
362. Searles K, Twksbury G, Tahmasebi B, Huffstutter A, Harvey J (1998) 43rd Int SAMPE Symp Exhib, Book 1, p 796

363. Nguyen MN, Grosse MB (1992) IEEE Trans Components Hybrids Manufacturing Tech 15:964
364. Possart W, Dieckhoff S (1999) Int J Adh Adhes 19:425
365. Iwaya Y, Ikeda T (1992) Jpn Kokai Tokkyo Koho JP 04 351,625
366. Tsunemi H, Nakata T, Namura K (1994) Eur Pat Appl EP 581,268
367. Arnold FE Jr, Rodriguez AJ, Granville A, Lyon RE (1994) Future Fire Retard Mater, Appl Regul Fire Retard Chem Assoc (Fall Conf), p 175
368. Tesoro G (1996) Proc Int Conf Fire Saf 22:119
369. Sase S, Mizuno Y, Sugimura T, Negishi H (1999) Japan Kokai Tokkyo Koho JP 11 21,503
370. Sase S, Mizuno Y, Sugimura T, Negishi H (1999) Japan Kokai Tokkyo Koho JP 11 21,504
371. Das S (1998) Int SAMPE Tech Conf 30:127
372. Ikeguchi N (1998) Japan Kokai Tokkyo Koho JP 10 142,797
373. Ikeguchi N (1998) Japan Kokai Tokkyo Koho JP 10 212,431
374. Ooya K, Sayama N (1996) Jpn Kokai Tokkyo Koho JP 08 39,685 (96 39,685)
375. Gesang T, Hoeper R, Dieckhoff S, Schlett V, Possart W, Hennemann OD (1995) Thin Solid Films 264(2):194
376. Kusumura K, Hayashi T (1998) Japan Kokai Tokkyo Koho JP 10 265,943
377. Hefner RE Jr (1985) US Appl 794,370,04
378. Ohya K, Sayama (1993) Eur Pat Appl EP 554,902
379. Ooya K (1993) Jpn Kokai Tokkyo Koho JP 05 222,148
380. Kada H, Nozaki M (1996) Jpn Koki Tokkyo Koho JP 08 138,929
381. Burack JJ, Fang T, LaGrange JD (1992) US Pats 5,165,959 & 5,208,892
382. Fang T (1991) US Pat 5,045,364

Editor: Prof. Joanny
Received: May 2000

Radical Polymerization in Direct Mini-Emulsion Systems

Ignác Capek[1], Chorng-Shyan Chern[2]

[1] Correspondence author
Polymer Institute, Slovak Academy of Sciences, Dúbravská cesta 9, 842 36 Bratislava,
Slovak Republic
e-mail: upolign@savba.sk
[2] Department of Chemical Engineering, National Taiwan University of Science
and Technology, Taipei 106, Taiwan

Abstract. Polymerization in direct mini-emulsions is a relatively new polymerization technique which allows the preparation of submicron latex particles within the range 100<particle diameter<500 nm. This process involves the generation of a large population of submicron monomer droplets in water (termed the mini-emulsion) by intensive shear force with the aid of an adequate emulsifier and coemulsifier (or hydrophobe). These stable, homogenized monomer droplets have an extremely large surface area and, therefore, can compete effectively with the monomer-swollen micelles, if present, for the oligomeric radicals gen-

Advances in Polymer Science, Vol. 155
© Springer-Verlag Berlin Heidelberg 2001

erated in water. Monomer droplets may thus become the predominant loci for particle nucleation and polymerization. This article presents a review of the current literature in the field of radical polymerization of conventional monomers and surface-active monomers in direct mini-emulsion systems. Besides a short introduction into some kinetic aspects of radical polymerization in direct emulsion, mini-emulsion and micro-emulsion systems, we mainly focus on the particle nucleation mechanisms and the common and different features between the classical emulsion and finer mini-emulsion polymerization systems. The effects of the type and concentration of initiator, emulsifier, coemulsifier (hydrophobe) and monomer are evaluated. The influence of the complex formation, the close-packed structure, etc. within the mini-emulsion droplet surface layer on the colloidal stability, the nature of the surface film, and the radical entry process are summarized and discussed. These results show that the nature of the coemulsifier (hydrophobe), the weight ratio of emulsifier to coemulsifier, the mechanisms of sonification, the stability of monomer droplets, and the presence or absence of free micelles play a decisive role in the polymerization process.

Keywords. Mini-emulsion polymerization, Monomer droplets, Particle size, Emulsifier, Coemulsifier/Hydrophobe

Abbreviations

ACPA	4,4′-azobis(4-cyanopentanoic acid)
AIBN	2,2′-azobisisobutyronitrile
AMA	allyl methacrylate
AMBN	2,2′-azobis(2-methylbutyronitrile)
AN	acrylonitrile
APS	ammonium peroxodisulfate
BA	butyl acrylate
BMA	bornyl methacrylate
C_{co}	solubility of coemulsifier in water
CA	cetyl alcohol
CE	conventional emulsion
CTAB	cetyltrimethylammonium bromide
CTAC	cetyltrimethylammonium chloride
CMC	critical micellar concentration
CHP	cumene hydroperoxide
$d(D_m^3)/dt_e$	the rate of Ostwald ripening
D	average particle size of the reaction mixture comprising both the monomer droplets and latex particles
D_{co}	diffusivity of coemulsifier
D_f	diameter of final latex particles
D_{fh}	diffusivity of hydrocarbon
D_m	diameter of monomer droplets
D_p	diameter of polymer particles
DBP	dibenzoyl peroxide
DDM	dodecyl mercaptan
DDMA	dodecyl methacrylate
DMBDA	2,5-dihydroxy-1,4-benzenedisulfonic acid

DPPH	diphenylpicrylhydrazyl
DTBHQ	2,5-di-*tert*-butylhydroquinone
E'	storage modulus
E''	loss modulus
[E]	emulsifier concentration
EDTA	ethylenediaminetetracetic acid
G_i	Gibbs free energy
HD	hexadecane
HMA	hexyl methacrylate
HQ	hydroquinone
[I]	initiator concentration
k_{des}	exit rate coefficient
KPS	potassium peroxodisulfate
LPO	lauryl peroxide
LSW	Lifshitz-Slyozov-Wagner theory
M_n	number average molecular weight
M_w	weight average molecular weight
$[M]_{eq}$	the equilibrium monomer concentration in the particles
MA	methyl acrylate
ME	mini-emulsion
MI	micro-emulsion
MMA	methyl methacrylate
MSt	α-methylstyrene
n	average number of radicals per particle
N_d	number of monomer droplets
N_m	number of latex particles generated by monomer droplet nucleation
N_p	number of latex particles
N_{pe}	Peclet number
N_w	number of latex particles generated by homogeneous nucleation
$N_{p,f}/N_{m,i}$	the ratio of the final number of latex particles to the initial number of monomer droplets
NMA	nonyl methacrylate
NP_{40}	nonylphenol polyethoxylate with an average of 40 EO units per molecule
o/w	oil-in-water
pMSt	*p*-methylstyrene
P_{dye}	the weight percentage of dye ultimately incorporated into the latex particles
PMMA	poly(methyl methacrylate)
PSD	particle size distribution
PSt	polystyrene
PVAc	poly(vinyl acetate)
PVC	poly(vinyl chloride)
PVOH	poly(vinyl alcohol)

r	droplet radius
R_i	rate of initiation
R_p	rate of polymerization
R_{pi}	initial rate of polymerization
RMA	alkyl methacrylate
rpm	stirring rate
SDS	sodium dodecyl sulfate
SHS	sodium hexadecyl sulfate
SMA	stearyl methacrylate
SN	sodium nitrile
SPS	sodium peroxodisulfate
St	styrene
t_a	aging time
T_g	glass transition temperature
TMPTMA	trimethylolpropane trimethacrylate
Tween 20	poly(oxyethylene) sorbitan monopalmitate
V	velocity of monomer droplets
V_m	molar volume of monomer (oil phase)
VAc	vinyl acetate
VC	vinyl chloride
VD	vinyl n-decanoate
VEH	vinyl 2-ethylhexanoate
VH	vinyl hexanoate
VS	vinyl stearate
Φ_{co}	volume fraction of coemulsifier
Φ_h	volume fraction of hydrophobe
Φ_m	volume fraction of monomer
Φ_p	volume fraction of polymer
$m_{m/h}$	molar volume ratio of monomer to hydrophobe
$m_{m/p}$	molar volume ratio of monomer to polymer
$\chi_{m/h}$	interaction parameter between monomer and hydrophobe
$\chi_{m/p}$	interaction parameter between monomer and polymer
γ	surface tension
ρ	entry rate coefficient
σ	oil droplet-water interfacial tension

1
Introduction

1.1
Oil-in-Water (o/w) Emulsions

Oil and water are essentially not miscible and coexist as a water phase and an oil phase, with each phase saturated with a trace of immiscible components. A surface active agent (emulsifier or surfactant) is soluble in one or in both phases, but it forms a true molecular solution only at a very low concentration. A mixture of oil, water, and emulsifier can form a milky (coarse) or transparent (fine) dispersion. The resultant dispersion is an oil-in-water (o/w) emulsion when a water-soluble surfactant such as the anionic sodium dodecyl sulfate (SDS) or non-ionic polyethoxylated nonylphenol with an average of 40 ethylene oxide units per molecule (NP40) is used. When the surfactant concentration is above its critical micellar concentration (CMC), these emulsifier molecules aggregate with one another to form micelles.

The o/w emulsions are often classified into three types according to their physical properties (the monomer droplet or particle size, turbidity, viscosity, etc.):
1) conventional emulsions prepared by ca. 1–3 wt % emulsifier have an average droplet size greater than 500 nm, exhibit a milky appearance and tend to separate upon standing,
2) mini-emulsions stabilized by ca. 5 wt % of emulsifier and coemulsifier (hydrophobe) have a mini-emulsion droplet (minidroplet) size in the range of 100–500 nm, show an opaque or milky appearance, but are more stable than the conventional emulsions, and
3) micro-emulsions stabilized by ca. 15–30 wt % of emulsifier and coemulsifier show a micro-emulsion droplet (microdroplet) size in the range of 10–50 nm, have a transparent or translucent appearance and low viscosity.

A mixture comprising three or four components (water, oil, emulsifier and comulsifier) can form both the kinetically stable (mini-)emulsion and thermodynamically stable mini- or micro-emulsion.

In a conventional, kinetically stable emulsion, each of the molecular units of the micelle or droplet may not be in equilibrium with a saturated aqueous solution. In emulsion polymerization, monomer partitioning between the large monomer droplets and growing latex particles depends strongly on the interfacial tension and interaction between the polymer and monomer. Diffusion of monomer from the monomer droplets to the growing particles is rapid and the equilibrium monomer partitioning is presumably a valid assumption for the polymerization system. In a thermodynamically stable micro-emulsion, however, the internal oil and external water phases are in equilibrium with each other. The partitioning of monomer among the oil phase, interface, aqueous phase, and polymer pseudophase can be calculated using a Flory-Huggins type approach [1]. In the case of complete monomer/polymer segregation, the monomer concentration at the locus of polymerization will be the molar concentration of monomer in the cores of the non-polymerized mini- or microdroplets. In the case of uniform monomer/polymer mixing, the monomer concentration at the locus of polymerization then depends on the degree of conversion at which the polymer has been produced to absorb the available monomer.

Emulsifying oil by conventional surfactants can proceed only to a certain extent before reorganization of the spherical micelles into an anisotropic structure, which is required to emulsify the extra oil, occurs. The stability of classical emulsions can be improved significantly by using a higher emulsifier concentration and/or stronger shear force, reducing the difference in the densities of the oil and water phases, and increasing the viscosity of the continuous phase. In conventional emulsions, the monomer molecules in the smaller droplets possessing a higher chemical potential tend to dissolve into the aqueous phase, while the larger droplets can grow at the expense of these small droplets. The process involving such a diffusional degradation of monomer droplets is termed the Ostwald ripening effect [2, 3]. Thus, emulsions subject to the Ostwald ripening effect are not stable and undergo phase separation upon standing for a short period of time. However, in mini-emulsions the presence of a low molecular weight and water-insoluble compound such as hexadecane (HD, hydrophobe) and cetyl alcohol (CA, coemulsifier) in the monomer droplets retards diffusion of the monomer out of the droplets. There is a question of whether the oil molecules are transported to the growing monomer droplets simply through the aqueous phase [2, 3], or by droplet collisions [4] and how the micelle evolution is influenced by the presence of coemulsifier (or hydrophobe). The addition of alcohol (coemulsifier) decreases the dissociation of emulsifier and, thereby, larger emulsifier aggregates appear (the curvature of the film decreases). Ostwald ripening is connected with the diffusion of the monomer through the water phase. The presence of free micelles even in the fine emulsion indicates the possibility of transferring the monomer by micelles [5]. The transport of reactants via micelles is accepted in the double emulsion [6]. Thus, the droplet stability might be discussed in broader relationships, i.e., the droplet surface layer, the type of the continuous phase and additives, and the presence or absence of free micelles.

The thermodynamically stable micro- or mini-emulsions can solubilize more oil and still remain isotropic, even with a minimal increase in the droplet size. The formation of thermodynamically stable mini- or micro-emulsions depends on the specific interactions among the constituent molecules. If these interactions are not realized, neither intensive homogenization nor excessive emulsifier will produce the thermodynamically stable emulsions. Once the preparation condition is right, spontaneous formation of the thermodynamically stable emulsion product occurs and only minimal mixing is required [7]. The properties of thermodynamically stable micellar solutions are time independent. These micellar solutions are independent of the order of mixing, and they will return to the original state when subject to a small disturbance which is subsequently relaxed. Most of the mini-emulsions are kinetically stable, but under an ideal condition thermodynamically stable mini-emulsions can also be prepared. The fact that the droplet size increases continuously with the aging time is a manifestation of the thermodynamic instability. Incorporation of a small amount of coemulsifier or hydrophobe (ca. 1–10 wt %) into the recipe can effectively retard the diffusional degradation of monomer droplets and maintain the very small droplet size generated by homogenization. This small droplet size (or large total droplet surface area) means that most of the emulsifier species are adsorbed onto the droplet surface. Therefore, little free emulsifier is available for micelle formation in the subsequent emulsion polymerization. That is, nucleation and polymerization in these tiny monomer droplets may control the reaction kinetics.

The term "mini-emulsion" is related to the submicron monomer droplets containing a coemulsifier, i.e., a compound with a certain surface activity such as CA. However, the monomer droplets also can be stabilized against diffusional degradation by addition of a small amount of a hydrophobic aliphatic hydrocarbon (e.g., HD) and polymer (e.g., PSt or PMMA). This indicates that the sole hydrophobe provides the monomer droplets with the osmotic pressure effect and is capable of stabilizing the monomer droplets alone. Besides, the presence of polar groups in the coemulsifier or polymer (e.g., CA, PMMA, etc.) favors the formation of a close-packed structure or complex on the droplet surface. Thus, diffusional degradation of the monomer droplets can be retarded or even prevented by addition of a surface-active component (coemulsifier) or a monomer-soluble hydrophobe. This is because coemulsifiers (hydrophobes) lower the Gibbs free energy of the mini-emulsion droplets, thereby decreasing the driving force for diffusion of the monomer into the aqueous phase. The saturated or unsaturated coemulsifiers reported in the literature include the long-chain alcohols, mercaptans, and alkyl (meth)acrylates, and hydrophobes include low-molecular weight aliphatic alkanes, high-molecular weight polymers, oil-soluble initiators, and very hydrophobic additives such as dyes [8–23]. The term "mini-emulsion polymerization" is also commonly used for dispersed systems in which particle nucleation predominantly occurs in the submicron minidroplets stabilized against Ostwald ripening.

Mixing oil and water in the presence of an appropriate amount of emulsifier and coemulsifier (hydrophobe) in a high-shear mixer leads to formation of a fine o/w dispersion. The highly monomer-swollen droplets thus formed can reduce the transport of monomer between the different phases. Adequate stability of the mini-emulsion system can be achieved by using a medium emulsifier concentration in combination with a hydrophobic, low-molecular weight coemulsifier (e.g., HD and CA). Furthermore, the presence of low- (e.g., HD) or even high-molecular weight (e.g., PSt) hydrophobe increases the colloidal stability of mini-emulsion droplets. The penetration of polar reactants (undissociated ionic emulsifier, coemulsifier, reaction products, etc.) into the monomer droplets is expected to increase the stability of monomer droplets. The close-packing of emulsifier and coemulsifier in the minidroplet surface layer through the hydrophilic (between the polar groups of emulsifier/coemulsifier and water) and hydrophobic (between the hydrophobic chains of emulsifier/coemulsifier and cosolvent) interactions then greatly depresses the diffusion of monomer from small droplets into large ones (Ostwald ripening [2, 3]).

1.2
Polymerization in Micellar Systems

In conventional emulsion polymerization, the monomer is partitioned among the following phases:
1) monomer droplets,
2) monomer-swollen micelles,
3) monomer-swollen polymer particles (after particle nucleation), and
4) aqueous phase.

Two characteristics of the o/w micro-emulsion polymerization are not found in conventional emulsion polymerization:
1) neither monomer droplets nor micelles are present in the reaction mixture and
2) the colloidal system is transparent.

The high ratio of emulsifier to water ensures that a large fraction of the emulsifier is undissociated and located at the droplet surface or even inside the droplets. Two features of mini-emulsion polymerization are different from those of conventional emulsion polymerization:
1) the number of micelles is small or they are not present at all in the reaction mixture and
2) the population of active micelles which becomes the monomer-swollen polymer particles is quite small.

Although, emulsion and mini-emulsion polymerizations show many common features, the particle nucleation mechanism and transfer of reagents for these two polymerization techniques are very different. Conventional emulsion polymerization is started with a heterogeneous system comprising monomer droplets, monomer-swollen micelles, and free emulsifier in water. The initiation reaction involves a two-step process. It starts with decomposition of the initiator and growth of initiator radicals in the aqueous phase. Subsequently, the oligomeric radicals enter the micelles or monomer droplets and induce the particle nucleation and growth processes. Particle formation takes place early in the reaction via homogeneous nucleation in the aqueous phase or entry of free radicals into the monomer-swollen micelles and minidroplets. Radicals also can enter the gigantic monomer droplets, but this event is insignificant for classical emulsion polymerization because of the relatively small total droplet surface area available for capturing radicals. Particle nucleation stops or slows down significantly after the total particle surface area has become sufficiently large to adsorb most of the emulsifier molecules. The monomer conversion up to the point at which nucleation stops is quite different for the classical emulsion, mini-emulsion or micro-emulsion polymerization; this conversion in the increasing order is: classical emulsion polymerization<mini-emulsion polymerization <micro-emulsion polymerization. The major locus of polymerization thereafter is in the nucleated latex particles. The reagents (monomer, emulsifier, oil-soluble additives, etc.) must diffuse from the monomer droplets to the active monomer-swollen polymer particles to support the polymer reaction therein and this transport process is the most pronounced for the classical emulsion polymerization, less for mini-emulsion polymerization, and the least for micro-emulsion polymerization.

A conventional batch emulsion polymerization can be divided into three rate regions [24–27] (Scheme 1). Particle nucleation occurs during Region I and is usually complete at a very low level of monomer conversion (much below 5%). During the early stage of polymerization, most of the monomer is located in the relatively large monomer droplets (ca. 1 mm in diameter). Particle nucleation takes place when the initiator radicals formed in the aqueous phase grow via the propagation reaction with monomer and then enter the monomer-swollen micelles [25–27] or precipitate out of the aqueous phase and form particle nuclei which may undergo limited flocculation until a stable primary particle population has been achieved [28–32]. The extremely large total surface area of micelles as compared to monomer droplets favors the participation of micelles in the initial absorption of the water-borne radicals. Homogeneous nucleation in the aqueous phase will also occur for the recipe with a low emulsifier concentration and/or a monomer with a relatively high water solubility. Thus, monomer droplet nucleation is insignificant because of the very small surface area associated with the large monomer droplets. Region II involves the primary polymerization of the monomer within the monomer-swollen polymer particles. The de-

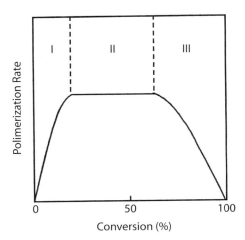

Scheme 1. The three rate regions involved in a conventional batch emulsion polymerization

pleted monomer in the growing latex particles is supplied by transport of monomer from the monomer reservoir (droplets) to the reaction loci (latex particles). Region III begins immediately after the disappearance of the monomer droplets and continues toward the end of polymerization (consumption of the remaining monomer in the polymer particles). In micro-emulsion polymerization, particle nucleation possibly proceeds throughout the reaction. In mini-emulsion polymerization, particle nucleation is shifted to a much higher conversion in comparison with the classical emulsion polymerization, but to a lower conversion in comparison with micro-emulsion polymerization. The mini- and micro-emulsion polymerizations do not exhibit all three intervals mentioned above.

From the viewpoints of the classical micellar and homogeneous nucleation models [25–32], the polymerization resulting exclusively from monomer droplet nucleation would have no Region II, in which a constant monomer concentration in the reaction loci is observed. Furthermore, there is no or only a minor dependence of the polymerization kinetics on the transport of monomer across the aqueous phase, which may be the rate-limiting step in some emulsion polymerization systems. Mini-emulsion polymerization refers to the reaction in the disperse system with a rather complex particle formation mechanism. The mechanism primarily involves monomer droplet nucleation and some contribution of micellar and homogeneous nucleations. The mini-emulsion polymerization technique involves dispersion of a large number of homogenized monomer droplets in water with the aid of emulsifier, coemulsifier, and cosolvents. These minidroplets may remain stable during polymerization by using an emulsifier in combination with a coemulsifier to depress diffusion of monomer from small droplets to large ones. The minidroplets have a very large total droplet surface area and, therefore, may compete effectively with the monomer-swollen micelles for the oligomeric radicals generated in water. Furthermore, the particle nuclei formed via homogeneous nucleation can be captured by these minidroplets. After capture of the oligomeric radicals from the aqueous phase, the polymerizing minidroplets are then transformed into the monomer-swollen polymer particles (latex particles). In addition, these nucleated droplets or highly monomer-swollen polymer particles might have a larger number of radicals per particle n than the monomer-swollen micelles. This is the reason why the polymerization kinetics is governed by pseudo-bulk kinetics rather than by the Smith-Ewart zero-one kinetics. In mini-emulsion polymerization, radicals enter the droplets and initiate the polymerization of monomer inside these droplets. The polymerization system under such a condition, especially with large minidroplets, might be regulated by the solution kinetics with participation of the gel effect [33]. In the early stage of polymerization, the nucleated minidroplets contain a large fraction of monomer and also may have a larger n due to the faster radical entry rate. As a result, the mini-emulsion polymerization rate is very fast. The properties of the latex product may be affected by the particle nucleation mechanism as well. The preferable formation of polymer inside the minidroplets would depress the interaction of radicals with emulsifier or chain transfer of radicals to emulsifier, which generally leads to incorporation of emulsifier units into the polymer matrix and changes in the physical properties of the latex product. The chemically incorporated emulsifier on the particle surface increases the colloidal stability of latex particles. The mini-emulsion polymerization, if other nucleation mechanisms are eliminated, might produce a latex product with a more uniform particle size distribution (PSD). However, the shrinkage of minidroplets with the progress of polymerization promotes the release of the originally adsorbed emulsifier due to which generation of a crop of particle nuclei in the aqueous phase may appear and the PSD may become broader.

Mini-emulsion polymerization using an effective emulsifier/coemulsifier system produces very small polymer particles. In this case, particle nucleation is primarily controlled by the radical entry into minidroplets, since very little emulsifier is present in the form of micelles or as free molecules in the aqueous phase. During the initial stage, the polymerization rate increases rapidly due to the increased concentration of particles (reaction loci). However, the particle formation process is suppressed by the minidroplet surface barrier to the entering radicals. This may result from the fact that the lifetime of the emulsifier at the minidroplet surface is increased. The high monomer concentration within the minidroplets promotes the propagation reaction and enhancement of the maximal polymerization rate. After the primary nucleation of minidroplets, the monomer level in these nucleated droplets continuously decreases during polymerization. As a consequence, there is no true Region II for the mini-emulsion polymerization system. The occasionally observed constant polymerization rate in the conversion range of 20–50% may result from the equilibrium between two opposing effects, that is, the growing population of latex particles and the decreased monomer concentration in the reaction loci. Besides, the gel effect (the Smith-Ewart case 3 or pseudo-bulk polymerization system) may counterbalance the continuously decreased monomer concentration in the polymer particles. A similar behavior can be found in micro-emulsion polymerization, where the dependence of the polymerization rate is described by a curve with one maximum rate at low conversion.

The size and number of monomer droplets play a key role in determining the locus of particle nucleation in emulsion, micro-emulsion, and mini-emulsion polymerizations. The total surface area of monomer-swollen micelles in the classical emulsion system is several orders of magnitude larger than that of monomer droplets. In the latter case, the micellar or homogeneous nucleation mechanism governs the overall particle formation process. After depletion of free micelles, the monomer-swollen latex particles predominantly capture oligomeric radicals from the aqueous phase. In micro-emulsion polymerization, the enormously large surface area of the micro-emulsion droplets (microdroplets) promotes the capture of oligomeric radicals from the aqueous phase, but disfavors the formation of radicals with a longer chain length and subsequent precipitation of the hydrophobic radicals out of the aqueous phase. The production rate of radicals is determined by the decomposition rate of the initiator, which can be taken as constant on the time scale of the polymerization. The assumptions of no water-phase termination, fast entry of radicals into microdroplets, and very high initiator efficiency (close to 100%) are generally accepted in micro-emulsion polymerization. Both the inactive microdroplets and monomer-swollen latex particles serve as an reservoir of monomer and emulsifier for the active particles (reaction loci). Besides, the nucleation of microdroplets leads to the release of emulsifier from the interface or core of microdroplets (undissociated emulsifier) and, therefore, increases the concentration of free emulsifier in the continuous phase. The competitive position of minidroplets for capturing radicals produced in water is enhanced by increasing the total droplet surface area and decreasing the amount of emulsifier available for formation of micelles or stabilization of primary particles generated by homogeneous nucleation. In addition, the rate of mini-emulsion polymerization practically decreases throughout the reaction. This is attributed to the continuous decrease of monomer concentration at the reaction loci due to polymer chain growth, dilution of the monomer phase by hydrophobic chains of emulsifier, undissociated emulsifier and coemulsifier, partitioning of monomer between minidroplets and inactive monomer-swollen polymer particles, and the chain-transfer of radicals to reactants, etc.

The chain transfer of radicals to monomer (or emulsifier) results in a dead polymer chain and a monomeric radical. The resultant monomeric radical may continue to propagate in the original particle. Alternatively, it may diffuse to another particle or monomer droplet by exit of this rather mobile radical to the aqueous phase followed by re-entry into another particle or by the intermediate droplet agglomerates in which monomer, emulsifier and radical are exchanged between the coalescing species. Whether the radical propagates in the same or a different particle is determined by the time scales of these events. To propagate within the original particle, the time scale of propagation must differ from the time scale of either monomer diffusion within the particle or a coalescence event [34]. Therefore, the probable fate of transferred radicals in micro-emulsion polymerization is exit and re-entry. In emulsion polymerization, exit of radicals is the most likely event for small particles [35]. However, the presence of the water-phase termination significantly decreases the initiator efficiency and the rate of classical emulsion polymerization. The competitive position of minidroplets for capturing radicals produced in water is enhanced by increasing the total droplet surface area but it decreases with the degree of close-packing of the droplet surface layer and of thickness of interfacial layer.

In mini-emulsion polymerization, the particle nucleation mechanism may be evaluated by the ratio of the final number of polymer particles to the initial number of monomer droplets ($N_{p,f}/N_{m,i}$). If the particle nucleation process is primarily governed by entry of radicals into the droplets, then the value of $N_{p,f}/N_{m,i}$ should be around 1. A lower value of $N_{p,f}/N_{m,i}$ may imply incomplete droplet nucleation or coalescence. On the other hand, a higher value of $N_{p,f}/N_{m,i}$ may indicate that the influence of micellar or homogeneous nucleation comes into play in the particle formation process, since one droplet feeds monomer to more than one micelle in the classical emulsion polymerization. For pure micellar nucleation (emulsion polymerization), the value of $N_{p,f}/N_{m,i}$ should be in the order of magnitude of 10^{2-3}. In micro-emulsion polymerization, on the other hand, the value of $N_{p,f}/N_{m,i}$ should be much less than 1.

Mini-emulsion polymerization can produce a latex product with a more uniform copolymer composition and PSD and better mechanical stability in comparison with the product obtained from conventional emulsion polymerization. The very large surface area of minidroplets ensures that the emulsifier is predominantly adsorbed on the polymer particle surface. This may thus improve the mechanical properties of the latex product. In addition, polymerization within the nucleated minidroplets, the reduced transport of monomer through the aqueous phase, and the side reactions can minimize the contamination of the latex product. One drawback of the current mini-emulsion polymerization using a volatile organic coemulsifier is the need to remove such a compound from the latex product. To overcome this problem, a surface-active co-monomer, an oil-soluble initiator, and a polymeric (oligomeric) hydrophobe were used as the sole coemulsifier to prepare kinetically stable mini-emulsions [13–22]. When an appropriate polymer is used as the coemulsifier (hydrophobe), the latex product can be used without further cleaning. Accumulation of a small amount of hydrophobic polymer within the nucleated minidroplets immediately after the start of polymerization is able to depress the diffusional degradation of minidroplets. This will then increase the number of latex particles and the colloidal stability of the disperse system. Indeed, the emulsion polymerization system initiated by an oil-soluble initiator promotes the accumulation of a small amount of polymer within the monomer droplets, leading to a decrease in the particle size as a result of the increased contribution of mini-emulsion polymerization [36]. In addition, accumulation of a small amount of polymer in a large number of monomer droplets can be reached by stopping the polymer chain growth. This can be achieved by incorporation of short-stopping agents or radical scavengers into the polymerization system.

Monomer droplet nucleation (mini-emulsion polymerization) suppresses the aqueous phase polymerization and, therefore, copolymerization of monomers with different water solubilities yields a copolymer that is rich in the water-insoluble component early in the polymerization [37]. The presence of highly water-insoluble additives (hydrophobes) decreases the monomer diffusion through the aqueous phase. In the classical emulsion copolymerization, however, the resultant copolymer is rich in the water-soluble monomer during the early stage of polymerization [38]. The homogenization step used for mini-emulsion polymerization produces submicron monomer droplets (ca. 100 nm in diameter), which are prevented from diffusional degradation by emulsifier/coemulsifier (or hydrophobe). As a result, these monomer droplets serve as the principal locus of particle formation in mini-emulsion polymerization. In the pseudo-conventional emulsion polymerization (with homogenization), the monomer droplets thus produced are very small initially, but they are quite unstable toward diffusional degradation. In such a polymerization system, the droplets should degrade immediately after homogenization, leading to a net decrease in the total droplet surface area. The resultant droplets should be larger than those of mini-emulsion, but smaller than those of conventional emulsion. In this case, the monomer-swollen micelles will be most likely the principal locus of particle nucleation. Nevertheless, the monomer droplets may play an important role in the particle formation process, depending upon their size at the time of addition of initiator.

Below the critical micellar concentration (CMC), micelles are not present in the mini-emulsion polymerization system and latex particles generated by homogeneous nucleation are fewer than those originating from minidroplets. However, it is not easy to stabilize the large oil-water interfacial area produced against the Ostwald ripening. To reach this goal, the PSD of minidroplets should be as narrow as possible. The diffusion of monomer from small droplets to large droplets is thus minimized. Furthermore, the emulsifier species should be immobilized in the droplet surface layer by the addition of suitable compounds. When homogenization is applied to a conventional emulsion system with the surfactant concentration above its CMC, the polymerization rate should be greater than that of mini-emulsion polymerization but lower than that of conventional emulsion polymerization in the absence of homogenization. This is because fewer micelles form in the reaction mixture subject to homogenization. The condition that the surfactant concentration is above its CMC in pure water does not necessarily imply that micelles are present in the mini-emulsion polymerization system, since these monomer droplets will have a much smaller size (stable enough against Ostwald ripening) and correspondingly a much larger total droplet surface area for emulsifier adsorption [39].

2
Formation of Monomeric Mini-Emulsions

2.1
General

Monomer emulsions tend to degrade by one of the following mechanisms. The first destabilization mechanism is coalescence of the interactive monomer droplets due to the attractive van der Waals force. This mechanism involves the rupture of the thin liquid film that forms between two adjacent droplets. This process requires the formation of a hole within the thin film which will then grow in size, resulting in the fusion of two colliding droplets. This coalescence process can be minimized by adequate emulsifier coverage on the droplet surface, which provides the colloidal system with sufficient electrostatic repulsion and/or steric repulsion forces to counteract the van der Waals force. For example, an anionic (or cationic) emulsifier can provide electrostatic repulsive force between two approaching colloidal particles [40, 41]. On the other hand, a non-ionic emulsifier can impart the particles with the steric stabilization mechanism [42, 43]. The second one is the Ostwald ripening process [2, 3]. This destabilization process refers to the diffusional degradation of droplets caused by transport of mono-

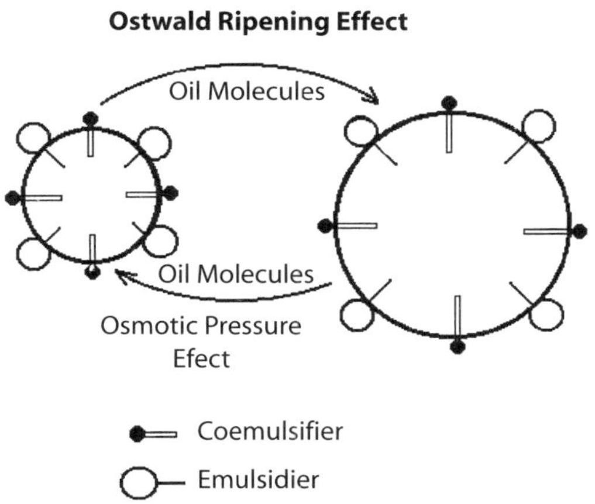

Ostwald Ripening Effect

Scheme 2. Mechanism of monomer transfer between two droplets

mer from the small droplets exhibiting a higher chemical potential, across the aqueous phase, and then to the large droplets (Scheme 2). Scheme 2 shows the mechanism for the transfer of monomer between two droplets with different sizes. Monomer tends to diffuse from the small droplet, through the aqueous phase, to the large droplet in order to relax the chemical potential gradient between two droplets with different sizes. This will then build up an osmotic pressure since the extremely hydrophobic coemulsifier cannot migrate from the small droplet to the large droplet to counterbalance the resultant concentration gradient between these two droplets. Thus, the monomer is forced to diffuse back to the small droplet and a relatively stable mini-emulsion is obtained. As a result, the total oil-water interfacial energy of the dispersed system is greatly reduced. This is the case even when the constituent components of the dispersed phase are only slightly soluble in the aqueous phase. Ostwald ripening results in an increase in the average droplet size and a reduction in the total oil-water interfacial area. Indeed, Jansson [44] showed that the disappearance rate of the degraded droplets can be very high for small droplets.

The extremely hydrophobic additives (hydrophobes) reduce the diffusion rate of monomer from the droplets to water, which then improves the shelf-life of these emulsion products. This is because diffusional degradation of the hydrophobe-containing monomer droplets will cause a gradient in the hydrophobe concentration between the small and large droplets. Since the hydrophobe molecules are incapable of diffusing out of the small droplets due to the rather limited water solubility, an osmotic pressure will build up to reverse the monomer transfer process between the small and large droplets (see Scheme 2). As a result, the colloidal stability of emulsion may be improved significantly by incorporation of a small amount of hydrophobe or coemulsifier (1–5 wt %) into the disperse phase, which effectively counterbalances the diffusional degradation effect. The hydrophobe used to prepare a monomeric mini-emulsion is highly water-insoluble and monomer-soluble in nature. Nevertheless, the hydrophobic coemulsifier may show some surface activity provided that a polar group is incorporated into its molecular structure. Typical examples of the hydrophobe and coemulsifier include HD and CA, respectively, and other types of extremely water-insoluble polymer, initiator, mercaptan, acrylic ester, dye, and mixtures of highly hydrophobic compounds have been employed to prepare relatively stable monomeric mini-emulsions [8–23]. A mini-emulsion stabilized by a long-chain alcohol such as CA is "thermodynamically" stable. On the other hand, a kinetically stable mini-emulsion, which undergoes appreciable diffusional degradation of monomer droplets upon aging, is generated when a less effective hydrophobe such as polymer is employed. In subsequent mini-emulsion polymerization, however, particle nucleation in the droplets can be completed before substantial diffusional degradation takes place [18, 19].

The preparation of a stable mini-emulsion (pre-emulsion) generally involves specially designed recipes and procedures. Such a homogenization process requires a relatively long pre-emulsification period with an intensive mixing and a total amount of ca. 5 wt % emulsifier and coemulsifier. In the case of a long-chain alcohol (e.g., CA, an amphiphilic component) with the gel formation, strong agitation is required to prepare a stable monomeric mini-emulsion. Otherwise, a mild agitation is usually sufficient to prepare a stable mini-emulsion [45]. When a long-chain alkane such as HD is used, a large shear force is required to prepare the stable mini-emulsion [46, 47].

The colloidal stability (or diffusional degradation) of monomer droplets can be further understood in terms of the Morton equation [48], in which the partial molar free energy of mixing of polymer in the monomer phase (spherical droplets) is expressed as:

$$\Delta G/RT = \ln \phi_m + (1 - m_{m/p}) + \chi_{m/p}\phi^2_p + 2\gamma V_m/r \cdot RT$$

where R is the universal gas constant, T is the temperature, ϕ_m and ϕ_p are the volume fractions of monomer and polymer, respectively, $m_{m/p}$ is the molar volume ratio of monomer to polymer, $\chi_{m/p}$ is the interaction parameter between monomer and polymer, γ is the oil-water interfacial tension, r is the droplet radius, and V_m is the molar volume of monomer. To define the thermodynamic conditions for equilibrium, the chemical potential of each monomer in each of the phases is required.

According to the Morton equation, increasing ϕ_p (or decreasing ϕ_m) reduces the magnitude of ΔG. The resistance against diffusional degradation of monomer droplets is therefore imparted to the disperse system. Incorporation of a polymer into the mini-emulsion recipe and/or formation of polymer inside the monomer droplets immediately after the start of polymerization allows the prolonged existence of the polymer-containing droplets in the reaction system. This is because the polymer concentration in the minidroplets should increase as the small droplets shrink due to Ostwald ripening. This diffusional degradation process is, however, thermodynamically unfavorable, as reflected by the increased osmotic pressure. Estimations based on the Morton equation show that the true colloidal stability achieved for the minidroplet size range generally requires a much larger amount of polymer in comparison with the experiments. The static mini-emulsion comprising a large population of tiny droplets is less susceptible to creaming (phase separation due to the gravitational force) in comparison with that containing huge droplets resulting from Ostwald ripening. Thus, the polymer-stabilized mini-emulsion technique can be used to prepare stable latex products.

In the presence of a hydrophobe, characterized by a relatively low molecular weight and water-solubility, monomer droplets are stabilized against diffusional degradation. The volume fraction of monomer in the hydrophobe-containing monomer droplets may be evaluated by the following equation:

$$\ln \phi_m + (1 - m_{m/h})\, \phi_h + \phi_h^2 \chi_{m/h} + 2V_m \gamma / r \cdot RT = 0$$

where ϕ_h is the volume fraction of hydrophobe in the particles and r is the droplet radius at the equilibrium swelling of droplets with monomer. The parameter $m_{m/h}$ is the ratio of the molar volumes of monomer to hydrophobe [49]. Thus, the degree of swelling of the hydrophobe-containing monomer droplets is primarily determined by the interaction parameter $\chi_{m/h}$ when a large value of $\chi_{m/h}$ is encountered. A major advantage of using a hydrophobe to stabilize the mini-emulsion product lies in the fact that the monomer droplet size may be controlled by the intensity of homogenization. In addition, the degree of swelling of the fine emulsion (monomer droplets) with monomer may be varied in a wide range.

The LSW theory dealing with Ostwald ripening [50, 51] is, strictly speaking, valid for the case of immobile oil droplets when the molecular diffusion is the only mechanism of mass transfer. Under these circumstances, the contributions of molecular and convective diffusion are related by the Peclet number (N_{pe}):

$$N_{pe} = r \cdot V / D_{fh}$$

where V is the velocity of droplets and D_{fh} is the diffusivity of hydrocarbon. Indeed, it was found that the Ostwald ripening rate data are systematically larger than the theoretical values [52]. An oil dispersion may degrade in many ways, for example, through aggregation, coalescence, creaming, and Ostwald ripening. Only coalescence and Ostwald ripening involve changes in the oil droplet size distribution and may result in complete phase separation. Ostwald ripening refers to the mass transfer between particles (or droplets) with different radii of curvature through the surrounding medium. The concentration of the dispersed phase material at the droplet surface is inversely related to the radius of curvature. Hence, a small droplet has a high surface concentration relative to a large droplet, thereby giving rise to a concentration gradient in the dispersed phase material. Mass is then transferred along the concentration gradient from the small droplets to the large ones. When Ostwald ripening occurs, small droplets tend to shrink and ultimately disappear, whereas large droplets grow at the expense of small droplets, eventually leading to phase separation. The LSW theory shows that, after a sufficient period of time, the Ostwald ripening process reaches a stationary state. It also predicts that the cube of the mean droplet radius increases linearly with the aging time. According to the LSW theory, the rate of Ostwald ripening $[d(D_m^3)/dt_a]$ for the mini-emulsion containing a water-insoluble, low-molecular weight coemulsifier can be predicted by the following equation [3]:

$$(d(D_m^3)/dt_a) = 64 \; \sigma \; D_{co} \; V_m \; C_{co} \; (\infty)/(9RT \; \phi_{co})$$

where t_a is the aging time, σ the oil droplet-water interfacial tension, D_m the average oil droplet size, D_{co} the molecular diffusivity of coemulsifier, V_m the molar volume of the oil phase, C_{co} (∞) the solubility of the bulk coemulsifier in water, and ϕ_{co} the volume fraction of coemulsifier in the oil droplet.

As mini-emulsion polymerization proceeds, polymer will form inside the monomer droplets. The nucleated droplet then consists of two phases, namely the polymer swollen with monomer and the hydrophobe, respectively. In such a case, the total swelling capability of monomer droplets is determined by both the polymer and hydrophobe (synergism or sum of the individual contributions). Experimental data obtained from mini-emulsion polymerization of vinyl chloride (VC) indicate that this effect is synergetic [53]. In this pioneering work, the seeded emulsion polymerization with a two-step particle swelling technique was used to prepare PVC particles. In the first step, the seed latex particles, which may be of any origin, are swollen with a highly compatible hydrophobic compound in a volume ratio from 1:1 to 5:1, and then swollen with monomer in the next step. It is very interesting to note that the presence of a hydrophobic compound in the seed particles increases the swelling capacity by several orders of magnitude as compared to the original seed particles.

A satisfactory stability of the pre-emulsion comprised of highly monomer-swollen particles can be achieved by using two different coemulsifiers or hydrophobes [e.g., a moderately water-soluble oil with relatively low molecular weight and viscosity (H1) and a highly water-insoluble oil with higher molecular weight and viscosity (H2)]. Under such a preparation condition (including homogenization), stable mini-emulsions can be obtained in a wide range of average droplet sizes and weight ratios (H1/H2) [54].

In the pre-emulsion stability studies, two different parameters are usually followed:
1) the rate of creaming, determined by the degree of separation of the milky emulsion from the transparent aqueous phase over a period of time and
2) the rate of droplet coalescence, determined by the amount of oil separated from the emulsion as a function of time.

The former is directly proportional to the square of the droplet size according to the Stokes-Einstein expression. In spite of the broad droplet size distribution and the changing droplet size with time, measuring the creaming rate can still provide a useful method for determining the relative droplet sizes of different emulsion products. The ultimate emulsion stability is not solely controlled by the diffusional degradation of oil droplets [55]. The presence of a liquid crystalline structure at the oil-water interface was referred to as the reason for the increased shear stability of the emulsified droplets. The studies of Choi et al. [56] and Ugelstad et al. [57] indicate that long-chain hydrocarbons and alcohols are favorable coemulsifiers (hydrophobes) for preparing stable monomeric mini-emulsions. El-Aasser et al. [58] reported that the success of mini-emulsion polymerization is dependent on the capability of producing the desired initial monomer droplet size distribution and the effectiveness of maintaining this distribution during the reaction. The most important factors in meeting these criteria are the molar ratio of emulsifier to coemulsifier and the means by which these amphiphiles are mixed. For example, in the preparation of mini-emulsions with HD, SDS is first dissolved in water and HD in the styrene (St) monomer [59]. The two solutions are mixed together and then homogenized to form the St mini-emulsion. A few extra steps are required in the preparation of mini-emulsions with CA. The aqueous solution of SDS and CA are subject to sonication in order to break up the gel phase thus formed. St is then added to the viscous solution and the mixture is homogenized to form the submicron minidroplets. The microfluidizer produces more uniform minidroplets than does the sonification equipment. This is because the microfluidizer provides the colloidal system with a more uniform exposure to the intensive shear force. Furthermore, the microfluidizer is suitable for the continuous production of mini-emulsion in industry.

When a significant quantity of monomer is contained in the droplets, there is a strong driving force for diffusion of monomer from small droplets to large droplets due to the Ostwald ripening effect. This monomer diffusion process is inherent in all emulsion systems because the small droplets have a higher chemical potential than the large droplets due to the variation of Gibbs free energy with droplet size. Thus, to minimize the free energy in the emulsion system, small droplets tend to disappear by diffusion of monomer from the small droplets into the large droplets. Indeed, Higuchi and Misra [60] showed that as the difference in the droplet size increases and an effective coemulsifier is not present in the emulsion system, the rate of degradation of small droplets increases significantly. In the emulsion system that is homogenized for a longer period of time this leads to the elimination of the population of large droplets. This then results in the most uniform and finest emulsion droplets. Therefore, diffusion of monomer from the small droplets to large droplets would be minimized. However, in the mini-emulsion subject to a shorter period of homogenization, there will be a significant degree of monomer diffusion from the small droplets to large droplets due to the presence of large droplets in the colloidal system.

2.2
Conventional Coemulsifiers

For a mini-emulsion prepared with a long-chain alcohol, the enhanced colloidal stability is attributed to both the retardation of diffusional degradation of monomer droplets and formation of the "intermolecular complex" at the oil-water interface. This interfacial complex film would be liquid condensed and electrically charged, thereby leading to a quite low interfacial tension and high resistance to droplet coalescence [61, 62]. However, Fowkes [63] ruled out the existence of the intermolecular complex formation in similar colloidal systems. It was concluded that the complex formation proposed for the pair of CA and cetyl sulfate does not prevent the emulsifier molecules from desorbing out of the droplet surface. That is, as the CA concentration increases, cetyl sulfate is removed from the interfacial zone. On the contrary, the work of Choi [64] and MacRitchie [65] suggested that the long-chain alcohol does prevent the emulsifier species from desorbing out of the droplet surface and covering the oil-water interfacial area which has not been occupied by emulsifier. As a result, an interfacial complex film, which resists rupture upon collision among the interactive droplets forms during the mini-emulsion preparation. CA has a hydroxy endgroup which forms a hydrogen bond with water. Thus, CA can be closely packed at the interspace between SDS molecules near the droplet surface layer. The resulting intermolecular complex at the oil-water interface then imparts a low interfacial tension and high resistance to droplet coalescence into mini-emulsion [66–68]. If a long-chain alcohol was used as the coemulsifier, the most stable mini-emulsion can be produced with a molar ratio of alcohol to emulsifier of between 2:1 and 3:1 [45, 69]. This result was then taken as an indication that some kind of close-packing or interaction between emulsifier and coemulsifier takes place and is favored at a certain ratio of emulsifier to coemulsifier. Azad et al. [70] also reported that there was an optimal ratio of emulsifier to coemulsifier for the best stabilization of a mini-emulsion (i.e., the smallest droplets) when CA was used as the coemulsifier. However, no such optimum has been observed in n-alkane-stabilized mini-emulsions. Since the thiol end-group in dodecyl mercaptan (DDM) exhibits the same chemical feature as a fatty alcohol, a similar oil-water interfacial behavior to the fatty alcohol-containing system was also observed in DDM-containing mini-emulsions [11,12]. The chain transfer agent DDM is capable of producing stable monomer droplets within the mini-emulsion droplet size, with the free emulsifier level below the CMC. The mini-emulsions stabilized by DDM are stable against creaming for at least three months. The Morton equation predicts the formation of stable minidroplets with a diameter of 200–400 nm for a certain level of DDM. At a constant level of emulsifier, increasing the DDM concentration results in a decrease in the monomer droplet size. Besides, the mini-emulsion products using DDM as the coemulsifier resist creaming for at least 17 h. The shelf-life of the mini-emulsion samples with the highest level of DDM reaches 3 months or longer.

The presence of a long-chain alcohol at the oil-water interface decreases the electrostatic repulsion force between the charged emulsifier molecules, which enhances the density of the droplet surface layer and probably promotes the formation of an interfacial complex film. The existence of the hydrophobic tail of coemulsifier retards the molecular diffusion of coemulsifier into the aqueous phase, but promotes the interaction between the hydrophobic tail of coemulsifier and the hydrophobic tail of emulsifier or the monomer phase. The interaction of the polar head groups of emulsifier and coemulsifier and the hydrophobic interaction within the oil-water interface can make the distinct interfacial film. Collisions between the interactive droplets do not lead to coalescence but rather to hopping (exchange of reactants).

A similar but more pronounced packing ability between emulsifier and coemulsifier was reported for the micro-emulsion system. A molar ratio of 1/1 for SDS/n-pentanol results in the smallest microdroplets, the maximal close-packed structure of emulsifier and coemulsifier, and the smallest surface coverage area of emulsifier at the oil-water interface [71]. An interesting approach demonstrated by Chern et al. [13–17] employed a reactive coemulsifier such as stearyl methacrylate (SMA) or dodecyl methacrylate (DMA) to prepare the St mini-emulsion. Just like conventional coemulsifiers, SMA or DMA acts as a coemulsifier in stabilizing the homogenized mini-emulsion. The steady-state value of the average monomer droplet size (D_m) for the St mini-emulsion aging at 35 °C decreases in the series:

CA (390 nm)>DMA (290 nm)>SMA (125 nm)~HD (115 nm)

Thus, the droplet size is inversely proportional to the hydrophobicity of the coemulsifier. Ostwald ripening occurring in the St mini-emulsion can be depressed by incorporation of additives such as DMA, SMA, CA, and HD [14]. The effectiveness of these coemulsifiers is strongly dependent on the coemulsifier solubilities in water. The solubilities of these coemulsifiers in water in the decreasing order were estimated to be:

CA (5.77×10^{-8})>DMA (1.38×10^{-8})>SMA (3.23×10^{-9})>HD (1.14×10^{-9} mL/mL)

The steady state values of D_m remain relatively constant (independent of the aging time) for the St mini-emulsions prepared by the hydrophobic coemulsifiers, HD and SMA. The average monomer droplet size, however, increases gradually and then levels off for the mini-emulsion using CA or DMA as the coemulsifier. This is attributed to the increased droplet degradation (Ostwald ripening) due to the initial broad droplet size distribution. In this case, the monomer molecules in the small droplets tend to diffuse through the aqueous phase and then into large droplets because the monomer solubility in water increases with decreasing droplet size. Judging from the data of the droplet size, creaming rate, and monomer phase separation, HD with the lowest solubility parameter is the best among the coemulsifiers studied. Alkyl methacrylates (DMA and SMA) are capable of producing relatively stable mini-emulsions. The performance of DMA is similar to CA, whereas the performance of SMA is similar to HD. Furthermore, the length of the alkyl group of the alkyl methacrylate influences the colloidal properties of the St mini-emulsion produced. When alkanes with a chain length of 10–16 carbons were used to prepare the mini-emulsion, no optimal ratio of alkane to emulsifier was observed (see above). However, Delgado et al., [72] found a plateau in the emulsifier adsorption experiments or good mini-emulsion stability could be achieved for a molar ratio of HD to sodium hexadecyl sulfate

(SHS) greater than 3 for the VAc and BA mini-emulsions. The presence of a small amount of HD causes a drastic increase in the amount of SHS adsorbed onto the monomer (VAc/BA=50/50 mol/mol) droplet surface (up to a molar ratio of HD/SHS=1/0.5). A further increase in the level of HD only results in a slight increase in the amount of adsorbed SHS. The amount of adsorbed SHS then levels off at a molar ratio of HD/SHS=3/1 [73]. For a given molar ratio of HD/SHS, increasing the HD concentration causes a decrease in the monomer droplet size with the result of extending the droplet surface area available for surfactant adsorption and, therefore, a sharp increase in the amount of adsorbed SHS is achieved. As expected, increasing the initial SHS concentration reduces the droplet size and, thereby, increases the amount of adsorbed SHS at any given molar ratio of HD/SHS. As the molar ratio of HD/SDS and/or the initial SHS concentration are increased, the droplet size is reduced and the stability of the resultant emulsion is improved. On the other hand, as far as the mini-emulsion stability is concerned, no optimal ratio was found for the molar ration of HD/SHS in the range of 1/1 to 10/1. Similar results were also observed for the St and MMA mini-emulsions with HD and SDS as the stabilizer pair [74], and for the VAc and methyl acrylate (MA) mini-emulsions stabilized by HD and Aerosol MA 80 [75].

2.3
Other Types of Hydrophobes

One problem associated with conventional coemulsifiers (e.g., CA, HD, etc.) is the need for removing these low molecular weight organic compounds from latex products. To overcome this drawback, a polymeric hydrophobe can be used as the sole coemulsifier to prepare kinetically stable mini-emulsions [18, 19, 76]. Polymers such as PSt and PMMA are highly water-insoluble but soluble in their own monomers. However, these polymers have a quite high molecular weight. The requirement that coemulsifier must be of low molecular weight is based on the swelling experiments and theoretical swelling calculation [77]. Experimental data reported by Miller et al. [76] and Shork et al. [18, 19] demonstrate that it is possible to produce mini-emulsions with the high molecular weight polymer as a poor coemulsifier. The polymeric hydrophobe delays Ostwald ripening to allow nucleation of monomer droplets by capturing oligomeric radicals from the aqueous phase. Once the droplets become nucleated, the polymer produced inside then adds additional stability against diffusional degradation. The emulsion products thus formed are not a true mini-emulsion in the sense that they are not stable over a period of months. Nevertheless, Ostwald ripening can be greatly reduced to permit the polymerization to take place in the minidroplets. The latex products obtained from the polymer-stabilized mini-emulsions exhibit all the characteristics of those obtained from the conventional coemulsifier-stabilized mini-emulsions. Moreover, most of the latex particles are derived from monomer droplet nucleation.

There is a range of PMMA molecular weight for preparing stable mini-emulsions [19]. This molecular weight range, which depends on both the emulsifier and PMMA concentrations, is from 3.5×10^5 to 7.5×10^5 g/mol. The mini-emulsion stability can be further enhanced by low emulsifier and high PMMA levels. The shelf-life of the mini-emulsion products ranges from 2 min to 6 h. The variables affecting the shelf-life of a mini-emulsion are the emulsifier and hydrophobe concentrations and the hydrophobe's (PMMA) molecular weight. The surface coverage of minidroplets by emulsifier increases with increasing polymer concentration and molecular weight of polymer. An optimal molecular weight is 3.5×10^5 g/mol. Since increasing the PMMA concentration results in smaller monomer droplets (or highly monomer-swollen polymer particles) and these droplets have a larger oil-water interfacial area, the droplet surface coverage goes down accordingly. A rise in the droplet surface coverage by one emulsifier molecule in the presence of polymer can be explained by the different polarities of monomer and polymer. The droplet surface area occupied by one emulsifier molecule increases with increasing polarity of the interfacial zone [78]. The monomer is more polar than the polymer; as the polymer concentration increases and more polymer is located near the droplet surface layer, the droplet surface area occupied by one emulsifier molecule must decrease. The droplet size decreases with increasing both the PMMA and emulsifier concentrations and varies in the range of 20–140 nm. The use of HD as the coemulsifier enhances the mini-emulsion stability significantly and, thereby, decreases the droplet size.

Wang and Schork [79] showed that a monomeric mini-emulsion can be prepared even with a poor polymeric coemulsifier. The authors used a non-ionic polymeric stabilizer (polyvinyl alcohol, PVOH), HD and/or polymeric hydrophobe (PMMA and PSt) to stabilize the mini-emulsion. The classical VAc emulsion containing PVOH is not stable. The shelf-life of the emulsion products is quite short and the average droplet size is generally larger than 2000 nm. The emulsion stability can be improved somewhat by using an additional stabilizer. Upon inncorporation of HD into the formulation, a true mini-emulsion is produced. Minidroplets have an average diameter of 300 nm or less and the shelf-life is varied in the range of 24–576 h. The data show a decrease in the droplet diameter (down to 57 nm) with increasing levels of stabilizer and hydrophobe. PMMA is a moderately good hydrophobe for preparing the MMA mini-emulsion. In an attempt to use PVAc (2800 g/mol) to make a VAc mini-emulsion, the pre-emulsions (highly monomer-swollen particles) are quite unstable, with a shelf-life of less than 1 h. This result is in agreement with the prediction of the Morton equation [48]. The theoretical calculation indicates that PVAc is almost totally ineffective in retarding Ostwald ripening in such a VAc emulsion. However, using PMMA as the hydrophobe, a marginally stable mini-emulsion can be obtained. The minidroplet diameter decreases from 420 to 262 nm when the PMMA concentration increases from 1 to 4% (based on total monomer weight). The shelf-life of the mini-emulsion products reaches 7.5 h. The presence of PMMA retards Ostwald ripening to the point where monomer droplet nucleation becomes dominant in the polymerization system. Under the same condition, the VAc mini-emulsion in the presence of PSt exhibits a larger droplet size and it is less stable in comparison with the PMMA system. These results indicate that not only the hydrophobicity and molecular weight of polymer but also other factors (such as incorporation of hydrophobe into the interfacial zone, interaction between polymer and emulsifier, compatibility between the seed and generated polymer, etc.) will have an influence on the mini-emulsion stability. The low activity of PSt as the hydrophobe seems to result from the weak interaction between PSt and emulsifier or the difficulty in the penetration of PSt chains into the oil-water interfacial zone. The presence of polar ester groups in PMMA implies the possible interaction between PMMA and emulsifier in the interfacial zone or orientation of PMMA chains near the droplet surface layer in such a way that an interfacial barrier to the monomer diffusion is established. The hydrophilic nature of PVOH or PVAc does not allow the penetration of polymer chains into the hydrophobic interior of the monomer/polymer particles and, therefore, PVOH or PVAc does not form an interfacial barrier.

The strong synergism in the formation of mini-emulsions was observed when PMMA and a highly hydrophobic alkyd resin were used as the hydrophobe [80]. The shelf-life of these mini-emulsion products varies from 7 to more than 50 days. The minidroplet size is below 300 nm. Based on the shelf-life and droplet size data, these mini-emulsions are somewhat between a kinetically stable mini-emulsion (with a shelf-life of a few hours only) and a truly stable mini-emulsion (with a shelf-life of a few months). In addition to PMMA, the presence of a water-insoluble alkyd resin evidently stabilizes the minidroplets against Ostwald ripening and forms a quite stable mini-emulsion. This may be due to the decreased solubility of the monomer or cosolvent in water and the increased adsorption of emulsifier on the droplet surface. The strong synergism in the formation of mini-emulsions was also observed in the systems containing PSt and CA [81]. When the homogenized emulsion does not contain predissolved polymer, the majority of monomer droplets disappears via Ostwald ripening. By contrast, the homogenized emulsion that contains the predissolved polymer exhibits a slower Ostwald ripening rate as compared to similar systems without the predissolved polymer. This enhanced colloidal stability is similar to what is noted for the mini-emulsion containing CA as the coemulsifier. Since both the homogenized emulsion and mini-emulsion systems (using CA as the coemulsifier) initially possess an unstable droplet size distribution, the increased stability enhancement in both systems was taken as evidence that the enhanced droplet nucleation caused by the predissoved polymer is primarily a result of preserving the droplet population by the predissolved polymer.

The effect of PSt on the St mini-emulsion in the presence or absence of CA was investigated by Miller et al. [82]. A solution of PSt in St was dispersed in the aqueous gel phase comprising SDS, CA, and water within the submicron droplet size range by the microfluidizer. The aqueous gel phase was prepared by mixing water, SDS, CA, and sodium bicarbonate and then homogenizing the resulting gel solution by ultrasonification. When polymer was used, PSt and St were mixed together until all of the polymer had dissolved in the monomer. The polymer solution was then added to the aqueous gel phase and mixed. The resulting emulsion was subjected to ultrasonification and then passed through the micro-fluidizer several times. The creaming rate for all of the emulsion products was measured by visually observing the phase boundary between the clear aqueous phase and the creaming emulsion phase. Assuming that the creaming rate is related to the droplet diameter, the droplet size of these emulsion products decreases in the series: (1% polymer, no CA)>>(1% polymer, CA)>(0.5% polymer, CA)>(0.05% polymer, CA)~(no polymer, CA).

The smallest droplets (i.e., the slowest creaming rate) are produced for the mini-emulsion in the absence of polymer, and the droplet size increases slightly with increasing polymer content (in the range of 0–1 wt %). In addition, the mini-emulsion prepared by 1% PSt and without CA phase separated very quickly, indicating that such a product is probably not stable toward diffusional degradation.

Stable St mini-emulsions prepared by using a water-insoluble, low-molecular weight blue dye as the sole coemulsifier were prepared successfully [23]. The Ostwald ripening rate for the mini-emulsions stabilized by SDS/dye is much faster (by several orders of magnitude) than those stabilized by SDS/DMA or SDS/SMA. This finding suggests that the blue dye within the concentration range investigated is not as effective as SMA or DMA in retarding the diffusional degradation of monomer droplets. One possible explanation for this difference is that the dye is less hydrophobic than CA or DMA. In addition, the molecular weight of the dye is much larger than that of CA or DMA. As a consequence, the complex formation or close-packing between emulsifier and dye near the droplet surface layer is not favored. The low coemulsifier activity of the dye may be attributed to its rigidity and bulkiness. These characteristics associated with the dye do not promote the formation of the close-packed structure at the oil-water interface. The interaction between hydrophobe and emulsifier within the interfacial layer, thus, is a measure of the hydrophobe activity.

In contrast to the electrostatic stabilization provided by SDS [40, 41], the non-ionic surfactant NP-40 imparts a steric repulsion force between two interactive hairy particles [42, 43]. Stable St mini-emulsions with NP-40 in combination with various coemulsifiers (CA, HD, DMA or SMA) were prepared and characterized [13]. The rate of Ostwald ripening for these mini-emulsions decreases in the series: CA>DMA>HD≡SMA.

2.4
Future Research on the Formation of Monomeric Mini-Emulsions

The water solubility of the hydrophobe is a function of temperature and the nature of the continuous aqueous phase. The water solubility of the hydrophobe decreases with decreasing temperature. Furthermore, the water solubility of the monomer is also lowered as the temperature is decreased. These factors may greatly suppress the Ostwald ripening effect. As a result, the colloidal stability of the emulsion may be improved significantly by lowering the temperature. A monomeric mini-emulsion is generally prepared at room temperature prior to polymerization. The resultant mini-emulsion is then heated to the polymerization temperature (ca. 60–80 °C for the thermally decomposed initiator), followed by addition of the initiator solution to start the reaction. This implies that the fundamental information about the mini-emulsion determined at room temperature may not be applicable to the mini-emulsion polymerization system. Addition of electrolyte varies the nature of the continuous phase as well as the degree of dissociation of the ionic emulsifier species. One example is the addition of the persulfate initiator to start the mini-emulsion polymerization. This might lead to the formation of a more close-packed structure of the surface active agent on the monomer droplet surface, thereby exhibiting an influence on the transport of the reacting species in the subsequent polymerization. The increased ionic strength may also have an effect on the colloidal stability of the mini-emulsion product according to the DLVO theory [40, 41]. Thus, determination of the colloidal properties of the monomeric mini-emulsion taking into account these considerations is required to gain a better understanding of the reaction mechanisms involved in the mini-emulsion polymerization.

Micelles comprising a mixture of both non-ionic and anionic emulsifiers are interesting because formation of such micelles is generally accompanied by a strong interaction between these two surface-active species. The decrease of diffusional degradation of the monomer droplets might be achieved by using a mixed emulsifier system, in which complex formation on the droplet surface layer might appear. The interaction between ionic and non-ionic emulsifiers results in the deviation from the ideal mixing. In the case of a positive deviation, the interaction between the hydrophobic chains of both emulsifiers is operative. When a non-ionic surfactant such as NP-40 is used, the hydrogen bonding between the polyethylene oxide part of NP-40 and water is very sensitive to changes in temperature. Thus, the interactive monomer droplets may coalesce with one another as a result of the greatly reduced mini-emulsion stability when the polymerization system is subject to the relatively high temperature. Moreover, the colloidal stability of the mini-emulsion stabilized by a mixed emulsifier system may be controlled by the ratio of the thickness of the non-ionic emulsifier adsorption layer to the thickness of the electric double layer of the droplets provided by the ionic emulsifier [83–85]. For example, the electrostatic repulsion force dominates the interparticle interaction process and, therefore, a relatively stable mini-emulsion is achieved provided that the electric double layer is much thicker than the non-ionic emulsifier adsorption layer. On the other hand, the mini-emulsion becomes unstable when the thickness of the electric double layer is comparable with the thickness of the non-ionic emulsifier adsorption layer. All these factors will make the task of developing a mechanistic model for the mini-emulsion polymerization stabilized by a mixed emulsifier system extremely difficult.

The high oil-solubility of a non-ionic emulsifier indicates that the non-ionic emulsifier may act as an ideal coemulsifier – it partitions between the aqueous and monomer phases. The Ostwald ripening effect is depressed when NP-40 is incorporated into the stabilization system (SDS/DMA) [15]. It is then speculated that this behavior can be interpreted in terms of the dissolved hydrophobe (emulsifier) or the presence of inverse micelles in the monomer phase (multiple emulsion). In the latter case, the presence of non-ionic and/or undissociated ionic emulsifier may lead to penetration of such polar molecules into the monomer droplets. The presence of emulsifier aggregates inside the monomer droplets may increase the stability of the multiple droplets and the total oil-water interfacial area.

Addition of a small amount of radical scavenger retards the propagation reaction of polymeric radicals in emulsion polymerization. The presence of a radical scavenger not only prolongs the particle nucleation period but also promotes the formation of a large number of highly monomer-swollen polymer particles. During the particle nucleation period, the polymeric molecules generated might be distributed among a large population of monomer droplets. Accumulation of a small amount of hydrophobic polymer inside the droplets retards the diffusion of monomer out of these highly monomer-swollen particles. Thus, monomer droplets contain not only the radical scavenger but also some less reactive unsaturated compounds, which would retard the radical growth events initially and, thereby, depress the transfer of monomer from the monomer droplets to polymer particles. Indeed, emulsion polymerization carried out in the presence of a small amount of a radical scavenger leads to the formation of latex particles with a smaller size [86, 87]. The enhanced stability of the monomeric mini-emulsion might be obtained by addition of a hydrophobic crosslinking agent, which decreases the growth rate and swelling ability of polymer particles. Accumulation of a small amount of microgel in the monomer droplets reduces the Ostwald ripening rate, diffusion of monomer to the reaction loci, and polymerization rate. Indeed, in the emulsion crosslinking polymerization, the rate of polymerization and particle size decrease with increasing level of crosslinking co-monomer in the reaction mixture [88,89]. These subjects are also of great interest to colloidal chemists.

According to the extended LSW theory, the rate of Ostwald ripening for a mini-emulsion upon aging is related to the monomer droplet-water interfacial tension and some physical properties of the coemulsifier such as the molecular diffusivity of the coemulsifier in water and the solubility of the bulk coemulsifier in water [3]. This theory may be useful in evaluating these colloidal parameters by measuring the Ostwald ripening rate. As an example, Chern and Chen used the LSW theory along with the data of the Ostwald ripening rate and monomer droplet-water interfacial tension for the St mini-emulsion to calculate the solubility of the reactive coemulsifier, DMA or SMA, in water. This approach is especially valuable for a coemulsifier whose water solubility is beyond the measurement limit of conventional instruments [14].

3
Kinetics of Mini-Emulsion Polymerization of Conventional Monomers

3.1
Effect of Fatty Alcohol and Mercaptan Coemulsifiers

The course of St mini-emulsion polymerization stabilized by SDS/CA can be divided into four major regions according to the maxima and minima observed in the polymerization rate (R_p) as a function of monomer conversion: R_p first rises to a primary maximum, falls off and exhibits an instability, increases to a secondary maximum again, and finally decreases rapidly toward the end of polymerization [90] (Scheme 3). The first maximal R_p is attributed to the formation and growth of polymer particles. The second maximal R_p is simply due to the gel effect. The monomer concentration in the reaction loci (latex particles) decreases with increasing conversion, and the monomer droplets may coexist with the growing particles throughout the polymerization. In the context of the micellar nucleation model [25–31], both the number of latex particles produced per unit volume of water (N_p) and the equilibrium monomer concentration or monomer concentration in the particles ($[M]_{eq}$ or $[M]_p$) contribute to changes in R_p with conversion. Therefore, the primary maximum does not necessarily correspond to the end of particle nucleation, since N_p can still increase with increasing conversion, but this effect can be outweighed by the contribution of $[M]_p$ to R_p. One of the key features of this reaction mechanism is a process whereby particle nucleation continues beyond the primary maximal R_p. In addition, it was suggested that radical entry into larger monomer droplets and polymer particles is favored over the smaller ones. As a result, a small number of large latex particles is nucleated early in the polymerization, and this population remains relatively constant throughout the reaction.

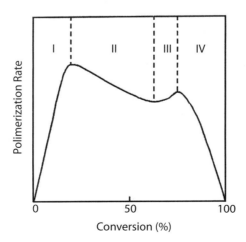

Scheme 3. The four rate regions involved in mini-emulsion polymerization

The experimental results of Miller et al. [90] suggest that mini-emulsion polymerization exhibits a long and slow particle nucleation. The length of this nucleation period is a function of the initiator concentration and the conversion at which nucleation ceases is in the range of 40–60%. As a result, the particle size distribution (PSD) is negatively skewed with a tail corresponding to the population of small particles. The degree of skew is dependent on the potassium peroxodisulfate concentration ([KPS]) (e.g., a lower initiator concentration produces a more skewed PSD). N_p increases very rapidly at the beginning of polymerization, leading to the situation that most of the latex particles are generated before about 10% conversion. At the beginning of polymerization, the rapid increase in N_p is due to the significant nucleation via the capture of the radicals by monomer droplets. Once a significant concentration of particles is generated, the formation rate of particles is reduced because most of the available free radicals are captured by the monomer-swollen particles. The exponent x=0.36 obtained for the relationship $N_p \propto [KPS]^x$ indicates that a certain fraction of monomer droplets initially present in the reaction system is nucleated and the population of nucleated droplets increases with increasing initiator concentration. This dependence may be attributed to the oil-water interfacial barrier formed by CA and SDS to the entering radicals. The exponent y=0.31 obtained from the relationship $R_p \propto [KPS]^y$, which is quite close to the above exponent x=0.36, implies that R_p is proportional to the number of reaction loci (particles) generated (Table 1). The values of both exponents may also vary with the decreased initiator efficiency with increasing [KPS], as shown by the work of Napper and Gilbert [91].

When the SDS concentration [SDS] is above its CMC, the overall polymerization rate of the conventional St emulsion polymerization is faster than that of the CA-containing mini-emulsion polymerization [59]. However, an opposite trend is observed when [SDS] is below its CMC. The R_p data agree with the final latex particle size data, that is, the faster polymerization rate corresponds to the smaller particle size (or the larger number of particles) produced. If the mini-emulsion is allowed to age before the start of polymerization, R_p decreases with increasing aging time. This is simply because the monomer droplet size increases with increasing aging time. The initial R_p of the mini-emulsion polymerization with CA as a coemulsifier is slower than that with HD. This might indicate that the radical entry rate and/or particle nucleation are depressed by the presence of CA in the oil-water interface (a close-packed structure of the interfacial layer). The PSD is narrower at the beginning of mini-emulsion polymerization and becomes broader or negatively skewed as the polymerization advances. This behavior is most likely due to the continuous monomer droplet nucleation. As the polymerization proceeds, the non-nucleated monomer droplets supply monomer by molecular diffusion to the growing particles. As a consequence, these minidroplets become smaller in size in the course of the polymerization. Once a radical does enter these droplets, the polymer is formed inside and these droplets are transformed into latex particles.

Since dodecyl mercaptan (DDM) shares some of the chemical features of a fatty alcohol such as CA, it is expected that DDM will exhibit some coemulsifier

Table 1. Kinetic parameters in the mini-emulsion polymerization initiated by KPS in the presence of SDS/classical coemulsifier[a]

Monomer	CoE[b]	Reaction order $(R_p \propto [I]^x, R_p \propto [E]^y, N_p \propto [I]^{x'}, N_p \propto [E]^{y'}, N_p \propto [CoE]^z)$					Reference
		x	y	x'	y'	z	
St, ME	CA	0.31		0.36			[90]
St, ME[c]	CA	0.2		0.2			[90]
St, ME	DDM		2.4				[11, 12]
St, ME	SMA	0.54		−0.2			[13, 15]
St, ME	DMA	−0.27				−0.35	[13, 15]
St, ME	SMA[d]	0.55		0.6			[112]
St, ME	DMA[d]	1.07	0.56	0.67	0.69		[112]
St, CE[e]		0.27		0.27			[119]
St, MI[e]		−0.45	1.5	0.32	1.3		[124]
St, CE	-	0.4					[110]
St, CE		0.4	0.6				[21, 22]
MMA, ME	HD	0.4	0.16	0.11	0.77	0.22	[108]
MMA, CE		0.39	0.24	0.28	0.51		[108]
MMA, ME	LPO	0.175	0.15	0.11	0.99[f], 0.15[g]	0.11	[22]
MMA, ME	VH			0.005			[110]
MMA, ME	DDM	-	–	0.06	-	0.36	[11, 12]
MMA, CE		0.2					[110]
BA/VAc, ME	HD	0.6	–	0.8	0.25		[16, 68, 92, 93]
BA/VAc, CE	–	0.36		indep.[h]	0.68		[16, 68, 92, 93]
VA, CE	–	–		indep.[h]			[100, 101]

[a] ME: mini-emulsion, CE: conventional emulsion, MI: micro-emulsion.
[b] CoE: coemulsifier.
[c] With AMBN.
[d] With AIBN
[e] Contribution of monomer droplet nucleation.
[f] Low [SDS].
[g] High [SDS].
[h] Independent

properties. Indeed, the polymerization of St or MMA in the presence of DDM is characterized by the mini-emulsion polymerization mechanism [11, 12]. For example, the final number of latex particles ($N_{p,f}$) is proportional to [KPS] to the power of 0.059 within the high initiator concentration range, which indicates that a large fraction of the monomer droplets is successfully nucleated. At a lower initiator concentration, not all droplets appear to be converted into latex particles. The change in the slope of the log $N_{p,f}$ vs. log [KPS] dependence occurs at a point at which effectively all droplets are nucleated, as determined by the ratio of $N_{m,i}/N_{p,f}$, where $N_{m,i}$ is the initial number of monomer droplets. In combination with SDS, DDM does not form a strong droplet surface barrier to the entering radicals as demonstrated by CA. The effective interaction of the entering radicals with DDM via the chain transfer reaction probably lowers the emulsifier/coemulsifier barrier for entering radicals (the high nucleation efficiency, $N_{p,f} \propto [KPS]^{0.059}$). In MMA mini-emulsion polymerization, the resultant latex particle size is quite close to the original monomer droplet size. This implies that droplets are the primary particle nucleation loci. The fact that $N_{m,i}/N_{p,f}$ is close to unity supports the mechanism of monomer droplet nucleation. The exponent x=0.36 obtained from the dependence $N_{p,f} \propto [DDM]^x$ indicates a quite high nucleation (coemulsifier) activity of DDM, in which [DDM] represents the DDM concentration in the monomer droplets (Table 1). This finding can also be attributed to the chain transfer reaction and slow propagation reaction. Accumulation of a small amount of polymer among a large number of monomer droplets and/or the short-stopping events promote the formation of a larger number of latex particles [86, 94]. Furthermore, the (short) polymer ending with a DDM unit is expected to exhibit some surface active properties.

The overall polymerization rate is faster for the classical emulsion polymerization in comparison with a mini-emulsion polymerization. This is because the emulsion polymerization system consistently develops smaller latex particles (i.e., a larger number of reaction loci) than the mini-emulsion counterpart. Nevertheless, the polymerization rate per particle ($R_p/N_{p,f}$) is faster for mini-emulsion polymerization. This is evidenced by the higher concentrations of radicals and monomer at the reaction site. Incorporation of DDM into the formulation decreases $R_p/N_{p,f}$ in both the conventional emulsion and mini-emulsion polymerization systems. This is due to the greater chain transfer (degradative) reaction, which enhances the radical desorption rate. Formation of polymer with a lower molecular weight supports the idea that DDM takes part in the chain transfer reaction. On the contrary, polymer produced in smaller latex particles generated via a micellar or homogeneous nucleation mechanism has a large molecular weight due to the lower DDM concentration in these reaction loci. For St mini-emulsion polymerization, the polymer molecular weight increases slowly with increasing conversion. For the classical St emulsion polymerization, however, the polymer molecular weight first increases very quickly up to 40% conversion, reaches a plateau (ca. 40–70% conversion), and then decreases slightly toward the end of polymerization. In the emulsion polymerization, DDM is consumed first in the nucleated monomer-swollen micelles. Since the diffusion rate of DDM from the monomer droplets to the growing particles is slower than that of St, the polymer molecular weight then begins to rise (Region II). The decrease of the polymer molecular weight observed in Region III is attributed to the consumption of the remaining DDM. Consumption of DDM in the minidroplets or polymer particles depends on the chain transfer constant and both the concentrations of DDM and monomer. Since DDM is uniformly distributed in the minidroplets (reaction loci), the polymer product obtained by mini-emulsion polymerization should have a lower molecular weight than that obtained by emulsion polymerization, where the DDM concentration at the locus of polymerization is mass transfer limited. The St mini-emulsion polymerization kinetics strongly varies with the SDS concentration ([SDS]), $R_p \propto [SDS]^{2.4}$, and deviates from the micellar nucleation model. This can be attributed to the increased total droplet surface area and stability of minidroplets and the immobilization of DDM within the oil-water interfacial layer (the hydration interaction between SDS and DDM). In the latter case, the influence of DDM on the propagation reaction decreases. Thus, an increase in [SDS] leads to an increased incorporation of DDM into the interfacial layer and decreased participation of DDM in the polymerization reaction. Indeed, the droplet size decreases with increasing [SDS] and [DDM]. The homogeneous nucleation cannot be ruled out because it would be favored when [SDS] increases. This nucleation route generates latex particles containing no DDM. Under these circumstances, a broad molecular weight distribution (or even a bimodal distribution) for the resultant emulsion polymer is expected.

3.2
Effect of Oil-Soluble Initiator Coemulsifiers

Lauryl peroxide (LPO) which acts as coemulsifier and initiator as well also shows some of the characteristics of a fatty alcohol. The role of LPO along with 2,2′-azobis(isobutyronitrile) (AIBN) and dibenzoyl peroxide (DBP) in the St mini-emulsion polymerization was evaluated in the literature [20]. The batch St mini-emulsion polymerization initiated by a series of initiators (AIBN, DBP, LPO) with different water solubilities led to the conclusion that LPO is capable of stabilizing the homogenized monomer droplets against their degradation by molecular diffusion. LPO is highly water-insoluble, highly monomer-soluble, and of low molecular weight. It meets all the criteria of a coemulsifier to prepare stable mini-emulsions. Several differences in the reaction kinetics for mini-emulsion polymerization with various oil-soluble initiators were also identified. For example, the average number of radicals per particle n is much larger for the polymerization with AIBN than that for the polymerization with LPO or DBP. Moreover, the smaller the latex particles, the larger is the difference in the reaction kinetics. This means that, in the polymerization initiated by AIBN, the latex particles can grow significantly during the reaction. On the other hand, in the case of LPO or DBP, R_p is extremely slow and, hence, the small latex particles will remain almost unchanged in dimensions in the course of the polymerization.

The polymerization rate may be parallel to the water-solubility of initiator and, therefore, the AIBN-initiated polymerization system is the most efficient. The water-phase initiation is much less pronounced in the polymerization system with DBP and it can be neglected in the system with LPO. The conformation of LPO near the droplet surface layer (LPO forms a close-packed structure with the emulsifier) depresses the entry of radicals from the aqueous phase to the latex particles. Immobilization of LPO in the oil-water interfacial zone probably depresses the initiator efficiency, as supported by the results obtained from the bulk polymerization in the high conversion region [95]. However, the comparable polymerization activity of LPO with DBP indicates that there must be some kind of mechanism responsible for formation of single-radicals in the particles containing LPO. This mechanism might involve the transfer of LPO by micelles, hopping events occurring during interparticle collisions, etc.

Theoretical calculations led to the conclusion that the probability of particle nucleation increases with an increase in the monomer droplet size. This means that large droplets are more easily nucleated as compared to small ones. This may explain why large latex particles appear early in the course of polymerization. For the mini-emulsion polymerization with LPO, the effect of compartmentalization of radicals on the polymer molecular weight distribution (MWD) is negligible and the molecular weight remains roughly constant at the beginning of polymerization and then increases significantly at high conversion due to the gel effect [21, 96]. Radical compartmentalization is also negligible for the mini-emulsion polymerization initiated by DBP, which is more water-soluble and less immobilized in the polymer matrix than LPO. However, a polymer with a rather low molecular weight is produced and, in spite of the conversion-time curve showing a distinct gel effect, the polymer molecular weight decreases as polymerization proceeds. According to the model developed by Alducin et al. [21], the reduction in the polymer molecular weight is due to the chain transfer to the initiator, which then counteracts the gel effect. It was then suggested that the DBP-saturated particles might act as a radical scavenger. The DBP molecules or radical pairs may react with the propagating radicals in the reaction volume in which the cage effect is important [97].

The molecular weight of the polymer produced by the homogenized St mini-emulsion initiated by AIBN (termed herein as mini-emulsion polymerization according to Ref. [21]) is larger than that obtained from the AIBN-initiated mini-emulsion polymerization in the presence of HD (termed herein as the classical mini-emulsion polymerization according to Ref. [21]). This difference is primarily due to the fact that in the classical mini-emulsion polymerization a significant fraction of polymer is produced in the small latex particles (80–200 nm), where substantial radical compartmentalization occurs. On the other hand, in the mini-emulsion polymerization, most of the polymer is formed in the large particles (>1000 nm) and negligible radical compartmentalization is experienced during the reaction. In both polymerization systems, the polymer molecular weight remains relatively constant at the beginning of polymerization and then increases sharply beyond 60% conversion due to the gel effect. Because of the more significant radical compartmentalization, the polymer molecular weight achieved in mini-emulsion polymerization is larger than that in the classical mini-emulsion polymerization. An important consequence of the radical compartmentalization is that the probability of bimolecular termination of radicals is greatly reduced. Therefore, R_p increases with an increase in the degree of radical compartmentalization.

At low concentrations of SDS and LPO, but still sufficient to form MMA mini-emulsion, the ratio $N_{p,f}/N_{m,i}$ is 0.5 [22]. Since all the monomer droplets contain some LPO and, therefore, these droplets are probably nucleated in the subsequent polymerization, the relatively low value of $N_{p,f}/N_{m,i}$ is attributed to the coalescence of droplets. At high SDS and low LPO concentrations, on the contrary, the ratio $N_{p,f}/N_{m,i}$ has a quite large value of ca. 29. Thus, latex particles may originate from monomer droplet, and micellar and homogeneous nucleation mechanisms. This assumption is also supported by an increase in the PSD of the latex product. The mini-emulsion polymerization with a high level of LPO gives a value of around unity for $N_{p,f}/N_{m,i}$. At low [SDS], the slope of the least-squares best fitted log $N_{p,f}$ vs. log [SDS] data is 0.99. At higher [SDS], the exponent x' in the relationship $N_{p,f} \propto [SDS]^{x'}$ then drops to 0.15 (Table 1). After the minidroplets are stabilized against coalescence, further increasing [SDS] does not greatly affect the droplet size, i.e., this action does not create new droplet surface area. The high emulsifier level ensures that the degree of dissociation of emulsifier is lowered. This can lead to an increase in the interaction between (co)emulsifier molecules, formation of a more dense oil-water interfacial structure and, as a result, hindering of radical entry. This postulation is supported by the fact that $R_p/N_{p,f}$ decreases with increasing [SDS]. Furthermore, the decreased monomer concentration at the reaction loci with increasing [SDS] also contributes to the decrease of $R_p/N_{p,f}$.

3.3
Effect of Alkane and Polymeric Hydrophobe

The high hydrophobe activity of HD was confirmed in the mini-emulsion copolymerization of vinyl acetate (VAc) and butyl acrylate (BA) stabilized by sodium hexadecyl sulfate (SHS) and initiated by KPS [23, 74, 98, 99]. The R_p vs. conversion curve shows four distinct regions (see Sect. 3.1). The maximal R_p is always attained at much higher conversion (ca. 20%) in mini-emulsion polymerization than that (ca. 10%) in the conventional emulsion polymerization. Relating region I primarily to particle nucleation, the experimental data then indicate that the particle formation rate in emulsion polymerization is much faster as compared to that in mini-emulsion polymerization. This is attributed to the reduced absorption rate of radicals by monomer droplets. Furthermore, the reduced formation rate of oligomeric radicals in mini-emulsion polymerization due to the suppressed droplet degradation, monomer transfer, and monomer concentration in the aqueous phase should also be considered. In the region II, R_p decreases with increasing conversion due to the preferential consumption of

BA and the continuous decrease in $\bar{n}$, which drops to a value less than 0.5 as a consequence of the increased rates of termination and radical desorption. R_p reaches the minimum at a conversion level in which BA has been exhausted. The end of region II is shifted to higher conversion as the fraction of BA in the recipe is increased. Region III corresponds to homopolymerization of the remaining VAc. The increased R_p in this region is a result of both larger propagation rate constant for VAc and an increase in $\bar{n}$ due to the gel effect. Finally, in region IV, R_p decreases with increasing conversion because of the depletion of monomer and the diffusion-controlled propagation reaction.

The dependence of the maximal R_p on [KPS] is stronger for mini-emulsion polymerization ($R_p \propto [KPS]^x$, x=0.6) as compared with the conventional emulsion polymerization {$R_p \propto [KPS]^{0.36}$ which is quite close to the micellar nucleation model (0.4)}, even though R_p is always faster in emulsion polymerization within the initiator concentration range studied. The exponent 0.6 is a result of two contributions:
1) N_p increases with increasing [KPS] and
2) the radical flux to a monomer droplet (or $\bar{n}$) increases with increasing [KPS].

Investigation of the effect of [KPS] on N_p in mini-emulsion polymerization yields the following dependence: $N_p \propto [KPS]^{x'}$, x'=0.8. In conventional emulsion polymerization, however, N_p was independent of [KPS]. Nomura et al. [100] and Friis and Nyhagen [101] reported similar results where N_p is independent of the initiator concentration for the conventional VAc emulsion polymerization using SDS as the emulsifier. The strong dependence of N_p on [KPS] in mini-emulsion polymerization indicates that only a small fraction of monomer droplets is nucleated at very low initiator concentration and the population of latex particles originating from monomer droplet nucleation increases significantly with increasing initiator concentration. Thus, increasing initiator concentration enhances the radical flux into droplets, the homogeneous nucleation and the concentration of surface-active agent (oligomer).

The influence of the emulsifier (SHS) concentration on N_p is more pronounced in the conventional emulsion polymerization system ($R_p \propto [SHS]^y$, y= 0.68) than in mini-emulsion polymerization (y=0.25). This result is caused by the different particle formation mechanism. While homogeneous nucleation is predominant in the conventional emulsion polymerization, monomer droplets become the main locus of particle nucleation in mini-emulsion polymerization. In the latter polymerization system, most of the emulsifier molecules are adsorbed on the monomer droplet surface and, consequently, a dense droplet surface structure forms. The probability of absorption of oligomeric radicals generated in the continuous phase by the emulsifier-saturated surface of minidroplets is low as is also the particle formation rate.

The presence of HD in the resultant BA/VAc latex particles obtained by mini-emulsion polymerization, in an amount far superior compared with that detected in the particles prepared in the absence of homogenization, provides strong evidence that particle nucleation occurs in the monomer droplets. When HD is used as the hydrophobe in mini-emulsion polymerization, it plays four important roles:

1) The presence of HD during the homogenization process promotes the formation and stabilization of submicron monomer droplets and most of the emulsifier species are adsorbed onto the droplet surface.
2) Upon initiating the polymer reaction, monomer droplets become the primary loci of particle nucleation. It is supposed that at this stage the adsorbed emulsifier and the generated polymer help the nucleated droplets retain the monomer originally contained therein.
3) Later on, HD in the uninitiated droplets reduces the equilibrium concentration of monomer in the growing polymer particles.
4) As a result of the nucleation in the HD containing monomer droplets, HD inside the polymer particles promotes the swelling capacity of polymer particles, as shown in the post-polymerization swelling experiments.

It is also interesting to note that HD may play two additional roles during mini-emulsion polymerization. It may hinder (or retard) the interparticle transport of monomer during polymerization by two different mechanisms:

1) Formation of an oil-water interfacial complex between HD and emulsifier by hydrophobic interaction, which may increase the interfacial resistance to entry (or exit) of species into (or from) the monomer droplets and
2) Retardation of the monomer diffusion due to the extremely low water solubility of HD, as suggested by Higuchi and Misra [60].

The droplet surface barrier to radical entry might also result from partial diffusional degradation of monomer droplets (monomer reservoir) and decrease in the droplet size. The reduction in the droplet size due to Ostwald ripening increases the density of the oil-water interfacial layer formed by emulsifier and HD. This will later lead to the formation of a strong barrier to radical entry.

In conventional emulsion polymerization, the disappearance of the VAc/BA droplets at ca. 25% conversion results from the transfer of monomer from monomer droplets to the locus of polymerization (monomer-swollen polymer particles). The presence of HD in the minidroplets reduces the free energy of mixing of the constituent monomers in the droplets. Therefore, the difference in the free energy of mixing between the monomer and polymer (particles) in mini-emulsion copolymerization is less than that in conventional emulsion polymerization. As a consequence, a smaller flux of monomer from the monomer droplets to polymer particles is achieved during mini-emulsion polymerization. In addition, HD cannot be transported from the droplets to particles because of its extremely low water solubility. Thus, the HD concentration in the droplets is greater than that in the particles, and monomer is retained in the droplets to minimize the HD concentration gradient.

It was shown that the basic principle of mini-emulsion polymerization can be extended to the reaction systems stabilized by cationic and non-ionic emulsifiers, leading to a narrow particle size distribution [102]. Besides, the effect of the initiator type [AIBN, KPS and V50 (a surface-active cationic initiator)] on the St mini-emulsion polymerization in the presence of HD was investigated. At similar molar concentrations of the cationic emulsifier, use of CTAB or CTAC (cetyltrimethylammonium bromide or chloride) results in latex particles of similar sizes. This implies that the particle size is essentially controlled by a limited emulsifier coverage of the latex particle surface. From surface tension measurement, this particle surface coverage was determined to be about 30%, which proves a very efficient use of emulsifier in the mini-emulsification process. A latex particle surface fully covered by emulsifier would allow emulsifier to form micelles in the water phase, and the interfacial energy of the water-air interface would reach the value of a saturated emulsifier solution. The relatively low surface tension then indicates that there are free micelles in the emulsifier solution. Two cationic emulsifiers (CTA terephthalate and tartrate), which were recently shown to be extremely efficient in the formation of micro-emulsions, exhibit only a moderate activity in mini-emulsification. Rather large latex particles with a close-to-complete emulsifier covered surface layer were obtained. This shows that the underlying energetic rules of micro- and mini-emulsions are quite different and the emulsifier efficiency relies heavily on the properties of the dispersed system. Non-ionic mini-emulsion can be made by using a PEO derivative (e.g., Lutensol, $C_{16-18}EO_{50}$) as the emulsifier, and the subsequent polymerization results in large, but also very well-defined, latex particles.

The reaction rate determined by calorimetry show insignificant dependence of the polymerization kinetics on the type of initiator or emulsifier. AIBN is mainly located in the monomer droplets while V50 is in the oil-water interfacial layer. However, the reaction temperature (70 °C) is high enough to promote formation of radicals by thermal initiation. This might suppress the effect of primary radicals derived from AIBN or V50. Addition of 1% polymer to the mini-emulsion stabilized by cationic emulsifier and initiated by V50 leads to an increase in R_p by a factor of 1.5. Under such conditions, polymer is not accumulated in the monomer droplets after the start of polymerization (homogeneous nucleation might take place). An accelerating effect with the predissolved polymer was not observed for the same polymerization system using AIBN as the initiator. The presence of AIBN in the monomer droplets is expected to produce PSt and, therefore, an additional (predissolved) polymer does not influence the reaction kinetics to an appreciable extent.

The combination of techniques of small-angle neutron scattering (SANS), conductivity and surface tension led to the conclusion that the primary monomer droplets in the St mini-emulsion polymerization stabilized by SDS and HD and initiated by AIBN have the same size and PSD as the final latex particles [103]. After stirring the St/HD/water mixture for one hour (no mini-emulsification), the conductance of the resultant emulsion is about 330 μS. After ultrasonification, the monomer droplet size decreases and the conductance further decreases to about 150 μS due to the significantly increased surface area of minidroplets. The conductance of the mini-emulsion product at 24 °C remains relatively constant, showing that the droplets are very stable during the monitoring time. This conductance level does not change on a time scale of days. Heating up the mini-emulsion with or without AIBN to the polymerization temperature (60 °C) leads to an increase in the conductance with the progress of polymerization and the increase is more pronounced in the run with AIBN. This was explained by the usual increase of ion mobility with temperature and additionally due to the lay-off of the single emulsifier molecule from the droplet surface at higher temperatures. At higher temperatures, the mini-emulsion product is also quite stable. It is interesting to note that the conductivity of the mini-emulsion remains constant at low temperature and then increases when the colloidal system is brought to the higher polymerization temperature. Under these circumstances, the thermal initiation is operative and monomer droplets can be saturated with St oligomer (Diels-Alder products [104]). This can contribute to the increased stability of minidroplets at high polymerization temperature. In the polymerization system with AIBN, where accumulation of polymer in the monomer droplets is more pronounced, the increase in the conductivity of mini-emulsion is stronger.

Latex particles with a narrow PSD (ca. 1.01) were achieved in the St mini-emulsion polymerization in the presence of CA initiated by the interfacial redox system of cumene hydroperoxide (CHP)/ferrous ion/ethylenediaminetetraacetic acid (EDTA)/sodium formaldehyde sulfoxylate [105]. Furthermore, the MWD of polymer produced varies in the range 1.5–2. The weight-average molecular weight (M_w) of PSt slightly decreases with increasing CHP concentration ([CHP]), while the initial rate of polymerization ($R_{p,i}$) increases very rapidly with increasing [CHP]. Both $R_{p,i}$ and limiting conversion decrease with increasing St concentration. This might be correlated with the partitioning of initiation redox components between the monomer and aqueous phases.

The polymer latex stability obtained from the mini-emulsion polymerization with various ratios of SDS/CA decreases in the series 1/3>1/10>1/1>1/6>1/0, which is consistent with the stability of monomer droplets reported by Ugelstad (1/3>1/2>1/1>1/6>1/0) [106]. The latex particle size decreases with increasing CA concentration. Furthermore, a two-dimensional hexagonal packing of surface-active molecules has been reported to be formed at a molar ratio of SDS/CA=1/3 in the colloidal system [107]. The good packing of the oil-water interfacial zone leads to satisfactory stability of monomer droplets, and it remains intact throughout the polymerization.

The dependence of the maximal R_p on [KPS] is quite similar for both the MMA mini-emulsion polymerization with HD (x=0.4) and the conventional emulsion polymerization (x=0.39) but different on [SDS] (y=0.16, ME) and (y=0.24, CE) [108]. The reaction orders x and y are a complex function of the radical entry (particle nucleation) and the extent of compartmentalization of radicals. The radical entry or particle nucleation increases R_p. N_p increases with increasing [KPS] and the degree of increase is more pronounced for the MMA emulsion polymerization ($N_p \propto [KPS]^{x'}$, x'=0.28) as compared with that for the MMA mini-emulsion polymerization (x'=0.11) (Table 1). The radical entry events are restricted due to the close-packed droplet surface layer, but the pseudo-bulk kinetics are more pronounced in mini-emulsion polymerization (the increased compartmentalization of radicals). Thus, the same reaction order x' or y' for both the emulsion and mini-emulsion polymerization systems does not imply the same reaction mechanism for both systems.

On the contrary, N_p increases more significantly with increasing [SDS] for the MMA mini-emulsion polymerization ($N_p \propto$ [SDS]$^{y'}$, $y'=0.77$) in comparison with the conventional emulsion polymerization ($y'=0.51$) (Table 1). The reaction order y' in mini-emulsion polymerization can also be discussed in terms of saturation or unsaturation of the droplet surface by emulsifier. In the former case under the emulsifier-saturated condition, the reaction order is relatively low. In the latter case (monomer droplet surface not saturated with emulsifier), the reaction order y' is large. The reaction order $y'=0.77$ indicates the emulsifier-starved condition experienced in mini-emulsion polymerization. Furthermore, the large value of y' also can be interpreted in terms of a higher hydrophobicity of the latex particle surface and increased adsorption of emulsifier. On the contrary, the dominant homogeneous nucleation produces polymer particles with a relatively low hydrophobicity (polymer ended with polar groups) and weak hydrophobic interaction with emulsifier. N_p increases with increasing HD concentration ([HD]) ($N_p \propto$ [HD]$^{0.22}$) as a result of the saturation of the droplet surface by emulsifier and/or increased droplet surface area.

Polymer (PMMA, PSt)-stabilized mini-emulsions were successfully prepared with the monomer droplet size in the submicron range [19, 79]. The majority of monomer droplets in the subsequent mini-emulsion polymerization is nucleated, i.e., the ratio $N_{p,f}/N_{m,i}$ ranges from 0.95 to 1.08. As a result of monomer droplet nucleation, the polymer-stabilized mini-emulsion polymerization is far less sensitive to variations in the recipe. The exponents for the dependencies of $N_{p,f}$ on the concentrations of initiator, water-phase retarder and oil-phase retarder and agitation speed are 0.002, 0.02, 0.0031, and −0.026, respectively. The corresponding values for the classical emulsion polymerization system are one to two orders of magnitude larger. PMMA has proven to be an excellent hydrophobe for the MMA mini-emulsion polymerization system. Nevertheless, PSt is less effective than PMMA in stabilizing the mini-emulsion products. The locus of particle nucleation is shifted from either the micelles or aqueous phase to the monomer droplets when polymer is incorporated into the polymerization system [19, 79, 81]. Such a shift to monomer droplet nucleation is a result of the homogenized droplets being preserved by the presence of the polymer. The homogenized emulsion that contains the predissolved polymer exhibits larger R_p and N_p as compared to a similar system in the absence of the predissolved polymer. This is similar to the mini-emulsion polymerization system using CA the coemulsifier. Since both the homogenized emulsion and mini-emulsion systems (using CA as the coemulsifier) initially possess a relatively unstable droplet size distribution, the enhancement of the reaction kinetics was taken as the evidence that the "enhanced droplet nucleation" is primarily a result of preserving the droplet population by the presence of polymer. Besides, the mini-emulsion polymerization (with PSt) shows a significant enhancement in R_p. The maximal R_p in the mini-emulsion polymerization containing the predissolved polymer is greater than that in the emulsion polymerization in the presence of the predissolved polymer. Above the CMC of the emulsifier solution, it was proposed that, in the homogenized emulsion without the predissolved polymer, nucleation predominantly takes place in micelles, although a limited degree of monomer droplet nucleation will also occur. The number of micelles is greatly reduced in the homogenized emulsion in comparison with the conventional emulsion since more emulsifier is adsorbed on the droplet surface. This phenomenon would explain the difference in the R_p data between the conventional and homogenized emulsions. In mini-emulsion polymerization, the locus of particle nucleation is shifted from the micelles to monomer droplets. Since the number of droplets in the mini-emulsion system is smaller than that of micelles in the homogenized or conventional emulsion systems, R_p would be lower in mini-emulsion polymerization.

The mini-emulsion polymerization stabilized by SDS, CA and PSt has a higher maximal R_p than does the homogenized emulsion polymerization containing PSt. This difference was discussed in terms of

1) reduction in the oil-water interfacial tension because the interaction between SDS and CA,
2) increased stability of monomer droplets with an interfacial barrier comprising SDS and CA, and
3) increased droplet surface area stabilized by SDS and CA.

Below the CMC, both monomer droplet nucleation and homogeneous nucleation take place in the mini-emulsion polymerization system. For the polymerizations carried out above and below the CMC, it was proposed that the dominant site for particle nucleation shifts to monomer droplets when PSt is incorporated into the homogenized monomer droplets. Both the mini-emulsion stabilized by CA and homogenized emulsion undergo monomer droplet degradation. However, the extent of diffusional degradation is quite different between these two colloidal systems. Both systems exhibit the "enhanced droplet nucleation", although to a different degree.

The mini-emulsion exhibits the highest value of R_p, the homogenized emulsion exhibits a slightly lower R_p, and the conventional emulsion shows the lowest R_p. The slowest polymerization kinetics associated with the conventional emulsion is typically ascribed to homogeneous nucleation. Since there is not enough emulsifier to form micelles, latex particles are generated by precipitation of oligomeric radicals in the aqueous phase. Because the water solubility of St is extremely low, homogeneous nucleation is a relatively inefficient process in the conventional St emulsion polymerization below the CMC. This then leads to the slowest polymerization kinetics. The homogenized emulsion exhibits a significantly faster polymerization as compared to the conventional emulsion polymerization system. Homogenization decreases the level of free emulsifier. Therefore, micelles are not present in the homogenized emulsion and stabilization of precursor particles generated by homogeneous nucleation is greatly lowered.

Thus, the major reason for the faster polymerization in the homogenized emulsion versus the conventional emulsion below the CMC is the presence of monomer droplet nucleation. Incorporation of polymer into the reaction system causes a significant increase in R_p. This is most likely due to the preservation of monomer droplets by a small quantity of polymer in each droplet produced by the homogenizer. The smallest latex particle size results from polymerization of the mini-emulsion containing the predissolved polymer (ca. 90 nm). The final latex particle diameter (ca. 110 nm) obtained from the homogenized emulsion containing the predissolved polymer is significantly smaller than that (ca. 200 nm) obtained from the homogenized emulsion in the absence of polymer.

As the ultrasonification time increases, both the rate of polymerization and the number of monomer droplets nucleated increase in the reaction system in the presence of the predissolved polymer [81]. This was taken as evidence that the mini-emulsion polymerization kinetics is directly related to the number of droplets. The R_p of the reaction system containing the predissolved polymer is always significantly higher than that in the absence of the predissolved polymer. A similar behavior was reported for the mini-emulsion prepared by the high shear microfluidizer. However, there was no significant increase in the rate of polymerization or number of monomer droplets nucleated brought about by incorporation of polymer into the mini-emulsion system when the low shear Omni Mixer was used for homogenization. The monomer droplets prepared by the low shear homogenizer were quite stable, whereas those created by the high shear homogenizer were relatively unstable. Since the enhanced droplet nucleation only occurs in the polymerization system with unstable droplets, this effect was then taken as supporting evidence that the primary root cause for the enhanced droplet nucleation is the preservation of monomer droplet number by the predissolved polymer prior to the polymerization.

As expected, the particle size of the latex product decreases with increasing ultrasonification time. Furthermore, the particle size for the sample containing polymer (1%) is significantly smaller than the counterpart without the predissolved polymer. These results are due to the fact that the number of monomer droplets available for polymerization is larger in these mini-emulsion systems. The dependence of the maximal R_p on [KPS] is weaker for the mini-emulsion polymerization with predissolved PSt (x=0.23) as compared with that for the mini-emulsion polymerization without PSt (x=0.33) [93, 109]. However, the maximal R_p for the mini-emulsion with predissolved PSt is much larger than that without PSt. In the first case, N_p is independent of [KPS] (x'=0.0). In the absence of the predissolved polymer, the particle nucleation increases with increasing [KPS] (x'=0.31) (Table 2). Thus, predissolving PSt in St monomer prior to the preparation of the CA-containing mini-emulsion results in a larger decrease in R_p and a reduction in the particle size as compared with a similar system in the absence of the predissolved polymer. Furthermore, the number of particles abruptly increases with predissolved PSt in the CA-containing mini-emulsion [76].

Table 2. Kinetic parameters in the mini-emulsion polymerization initiated by KPS in the presence of SDS/non-conventional coemulsifier[a]

Monomer	X	Reaction order		Additive	Ref.
		$R_p \propto [I]^x$	$N_p \propto [I]^{x'}$		
		x	x'		
St, ME	KPS	0.23	0.31		[93, 109]
St, ME	KPS	0.33	indep.	PSt	[93, 109]
MMA, ME	KPS		0.002	PMMA	[19, 79]
MMA, ME	KPS/NS		0.02	PMMA	[19, 79]
MMA, CE	KPS/NS		0.153		[19, 79]
MMA, ME	KPS/DPPH		0.003	PMMA	[19, 79]
MMA, CE	KPS/DPPH		0.153		[19, 79]
BA/AN, CE	KPS/DPPH		0.1		[86]
MMA, ME	rpm[b]		−0.026		[19, 79]

[a]See legend to Table 1, X: initiator or initiator/inhibitor.
[b]Stirring rate.

The presence of a small amount of polymer (PSt or PMMA) inside the homogenized monomer droplets reduces the sensitivity of N_p to changes in [initiator]$^{x'}$ and the decrease of the dependency is more pronounced for the mini-emulsion polymerization with PMMA (Table 2) [19, 79, 93]. Under these circumstances, the added polymer increases the lifetime of monomer droplets and the probability of monomer droplet nucleation. The ratio $N_{p,f}/N_{m,i}$ was found to be very close to 1 for the mini-emulsion polymerization with PMMA, but it is much above 1 for the run with PSt. The interaction between the polymer particle surface and emulsifier increases with increasing hydrophobicity of polymer and, thence, PSt should promote the formation of more stable monomer droplets in the preparation of the mini-emulsion. However, the reverse seems to be true for the highly diluted polymer particles with the predissolved PSt. PSt mainly locates in the monomer droplet core, whereas the more hydrophilic PMMA tends to diffuse closer to the droplet surface layer and interact with emulsifier therein. Thus, the stronger interaction of PMMA with the droplet surface in comparison with PSt makes PMMA a more efficient hydrophobe.

It is known that changing the molecular weight of a polymer affects the bulk viscosity of the dilute polymer solution. In addition, modifying the polymer chain end will change the close packed structure of the CA/SDS complex at the oil-water interface. However, predissolving PSt without or with SO_3^- end-groups (a number-average molecular weight ranging from 3.9×10^4 to 2.06×10^5 g/mol) in the St monomer results in the same enhancement in the polymerization kinetics. This is then taken as evidence against either a droplet surface modification or a change in the droplet viscosity which are the determining factors in the "enhanced droplet nucleation." However, the monomer droplet size data in the absence of polymer indicate that these droplets undergo diffusional degradation and, therefore, the average droplet size increases with the aging time. It is then suggested that the predominant cause of the "enhanced droplet nucleation" is that the addition of polymer contributes to the stability of the mini-emulsion droplets both prior to and during the polymerization. This can be primarily attributed to the preservation of droplets by incorporation of polymer into the mini-emulsion droplets produced during homogenization. Thus, polymer helps the colloidal system preserve monomer droplets for a longer period of time during polymerization as compared to the conventional mini-emulsion. Polymer is unable to preserve the size of monomer droplets produced, but instead only the number of droplets produced during homogenization [92].

Addition of CA to the mini-emulsion system reduces the oil-water interfacial tension. Increasing the PSt (without SO_3^-) concentration, however, has little effect on the interfacial tension. This is because these polymer chains are not capable of penetrating into the interfacial layer and, thus, do not affect the interfacial tension. Increasing the level of polymer with an SO_3^- end-group slightly reduces the interfacial tension. The PSD of the latex particles increases with increasing conversion, and at higher conversion there exists a distinct bimodal distribution. This bimodal distribution was attributed to two different phenomena [93]. The larger latex particles were produced by nucleation and growth of the mini-emulsion droplets. The smaller particles were then generated by diffusion of monomer out of the original droplets containing the predissolved polymer and CA. It is expected that these small droplets contain a large quantity of CA.

The overall activation energy (E_o) for the mini-emulsion polymerization process was estimated to be 52.6 kJ·mol^{-1}. E_o decreases with increasing temperature ($E_{o,60-70\,°C} > 52.6$ kJ·mol^{-1} and $E_{o,80-90\,°C} < 52.6$ kJ·mol^{-1}). In addition, the limiting conversion of the polymerization system increases from 76 to 92% when temperature increases from 60 to 90 °C. This can be discussed in terms of the thermal initiation of styrene which increases with increasing temperature. The limiting conversion might result from the accumulation of CA and St in the inactive micelles. Indeed, the strong synergism in the formation of the mini-emulsion containing PSt and CA [81] might be responsible for the preservation of the inactive CA/St swollen micelles or highly monomer-swollen polymer particles. Furthermore, the contribution of polymerization in the glassy state is not ruled out and decreases with increasing temperature.

The highly hydrophobic alkyd resin can serve as both reactant and coemulsifier (hydrophobe) in the mini-emulsion polymerization of alkyl (meth)acrylates [80, 110]. The strong synergism in the minidroplet nucleation was observed when highly hydrophobic unsaturated alkyd resin was used as the hydrophobe (≥20 wt %). This feature is very different from the conventional coemulsifiers, such as long-chain alkanes and alcohols, that are effective around at 2 to 4%.

Furthermore, this work demonstrated that the mini-emulsion technique provides an ability to form graft copolymer. Resin shows the inhibitory effect on the rate of polymerization. This can be connected with the slow accumulation of polymer within the monomer droplets and the increased stability of minidroplets. Indeed, the polymers were relatively small. The polymers obtained had a wide molecular weight distribution, with a polydispersity of more than 19 and M_n slightly larger than that for alkyd. Alkyd resins contain multiple double bonds located at different positions along the fatty acid backbone. Approximately 20% of the double bonds in alkyd are consumed in grafting reactions. Three glass-transition temperatures were observed, indicating the presence of at least three forms of polymer. The two higher glass-transition temperatures correspond to the calculated T_gs of poly(acrylate-graft-alkyd) and polyacrylate, respectively. The proportion of the two kinds of polymers in the samples was determined by the extraction and indicates that graft copolymer was the predominant form.

3.4
Effect of Reactive Coemulsifiers

Water-insoluble, unsaturated co-monomers such as p-methylstyrene (pMSt), vinyl hexanoate (VH), vinyl 2-ethylhexanoate (VEH), vinyl n-decanoate (VD), and vinyl stearate (VS) are able to impart stability to the mini-emulsion droplets against diffusional degradation [111]. These extremely hydrophobic co-monomers, like the conventional coemulsifier (or hydrophobe) such as CA or HD, can eliminate the diffusional degradation of monomer droplets and produce droplet sizes within the mini-emulsion range. Furthermore, the carbon-carbon double bond can be chemically incorporated into latex particles in the subsequent polymerization. Therefore, these organic compounds show no negative effects on the emulsion polymer products. These submicron droplets eliminate or compete effectively with micellar and homogeneous nucleation. The majority of monomer droplets is nucleated, as reflected by the ratio of $N_{p,f}/N_{m,i}$ ranging from 2.3 to 5.5. $N_{p,f}$ is almost independent of [KPS] in the mini-emulsion copolymerization of MMA and VH ($x'=0.005$). Emulsion polymerization, in contrast, shows an exponent of $x'=0.2$ and 0.4 for the reaction system containing MMA and St, respectively.

Dodecyl methacrylate (DMA) and stearyl methacrylate (SMA) are efficiently reactive coemulsifiers in the St mini-emulsion polymerization using SDS as the emulsifier [13]. The high coemulsifier (hydrophobe) activity of these unsaturated compounds results from the fact that the R_p data are almost the same in the mini-emulsion polymerizations stabilized by SDS/CA, SDS/DMA, SDS/SMA and SDS/HD. The average particle size (D) of the reaction mixture comprising both the monomer droplets and latex particles in the SDS/CA or SDS/DMA stabilized polymerization decreases rapidly from ca. 300–350 nm to a minimum (ca. 115–125 nm), followed by a gradual increase to a plateau (ca. 125–150 nm). The initial abrupt decrease in D is attributed to the generation of particle nuclei by homogeneous nucleation [29–31], even though [SDS] is only slightly above its CMC. The mini-emulsion with SDS/CA or SDS/DMA undergoes significant diffusional degradation, which results in a decrease in the total droplet surface area and the release of some of the adsorbed SDS molecules. The desorbed SDS molecules can stabilize the primary particles formed by homogeneous nucleation. A similar trend in the D vs. time dependence was also observed in the SDS/SMA and SDS/HD stabilized mini-emulsion systems, but the initial decrease in D is not significant (i.e., D decreases from ca. 150 to 125 nm only). This is due to the lack of stabilizing species to induce the population of very small primary particles and monomer droplet nucleation is the predominant mechanism for producing latex particles. The predominant homogeneous nucleation in the polymerizations with SDS/CA and SDS/DMA is further supported by the fact that the ratio $N_{p,f}/N_{m,i}$ is greater than unity. On the other hand, the value of $N_{p,f}/N_{m,i}$ is ca. 0.8 for the polymerization with SDS/SMA or SDS/HD:

$N_{p,f}/N_{m,i}$: 19.2 (CA)>7.2 (DMA)>>0.86 (HD)>0.76 (SMA).

It was found that the SDS/CA containing polymerization system shows a bimodal PSD of latex particles, and the SDS/DMA containing system is characterized by a quite broad PSD. The fraction of small latex particles results from the initially generated highly monomer-swollen polymer particles, which serve as the monomer reservoir in the later stage of polymerization. On the contrary, the latex products obtained from both the SDS/SMA and SDS/HD containing systems show a relatively narrow PSD. These data further support the proposed competitive particle nucleation mechanism.

The reagent DMA or SMA forms a close-packed structure with NP40 on the monomer droplet surface layer [14]. In contrast to the SDS stabilized mini-emulsion polymerization, R_p for the runs with NP40 increases in the following series:

CA<DMA<HD<SMA.

This trend correlates reasonably well with the $N_{p,f}$ data; the larger the value of $N_{p,f}$, the greater is the value of R_p. During the very early stage of polymerization, the rapid decrease in D was also observed in these NP40 stabilized polymerizations. The magnitude of reduction in D increases in the following order:

HD<<DMA<SMA<CA.

The initial decrease in D might be attributed to the accumulation of hydrophobic oligomer in the monomer droplets due to the chain transfer of free radicals to NP40 dissolved in the monomer phase. In the case of highly stable mini-emulsion droplets saturated with HD, the accumulation of hydrophobic oligomer in these droplets does not affect the polymerization kinetics synergistically. The variation in D is also strongly dependent on the polymerization temperature. This can be attributed to the reduced strength of hydrogen bonds between the PEO part of NP40 and water, thereby lowering the hydrophilicity of NP40. This may then impair the steric stabilization effect provided by NP40 and cause coalescence among the interactive monomer droplets. Mini-emulsions prepared at temperatures much higher than 60 °C show significant diffusional degradation of the monomer droplets (or droplet coalescence). This results in a reduction of the total droplet surface area and, consequently, the release of some NP40 species because the droplet surface is overcrowded with NP40. Thus, the desorbed emulsifier molecules may contribute to the stabilization of primary particles generated in the aqueous phase. The presence of NP40 in the oil phase and the release of the absorbed NP40 during polymerization also promote the delayed (secondary) particle nucleation [112].

The contribution of homogeneous nucleation is supported by the $N_{p,f}/N_{m,i}$ ratio, which is much greater than unity for the mini-emulsion polymerization with CA or DMA. On the other hand, $N_{p,f}/N_{m,i}$ is close to unity for the polymerization with HD or SMA. The magnitude of $N_{p,f}/N_{m,i}$ for the mini-emulsion polymerization with various coemulsifiers or hydrophobes decreases in the series:

$N_{p,f}/N_{m,i}$: 19.6 (DMA)>18.1 (CA)>3.7 (SMA)>1.1 (HD).

Thus, the polymerization system with CA or DMA shows that a large proportion of the resultant latex particles with a quite broad PSD are produced by homogeneous nucleation. On the other hand, the polymerization system with HD or SMA includes the predominant monomer droplet nucleation, leading to a narrow PSD for the latex product. The mixed particle nucleation mechanism was illustrated in the SDS/DMA stabilized polymerization by using the blue dye as a probe [15]. It was shown that approximately 55% of the initial monomer droplets is transformed into latex particles and the number of latex particles originating from the monomer droplets (N_m) increases with increasing DMA concentration ([DMA]). An opposite trend was observed for the number of latex particles originating from the particle nuclei generated in water (N_w). Thus, the degree of homogeneous nucleation and R_p decrease with increasing [DMA], as confirmed by the following relationships (Table 1):

$R_p \propto [DMA]^{-0.27}$, $N_{p,f} \propto [DMA]^{-0.35}$, $N_m \propto [DMA]^{0.95}$, and $N_w \propto [DMA]^{-0.43}$.

The initial monomer droplet size ($D_{m,i}$) for the SMA containing mini-emulsion is much smaller than that for the DMA counterpart, owing to the more effective SMA in retarding diffusional degradation of the monomer droplets. Analysis of the particle nucleation data led to the conclusion that in the SMA containing mini-emulsion polymerization a large fraction of monomer droplets can be converted into latex particles. However, a relative large population of latex particles is still produced via homogeneous nucleation. In support of this conclusion are the variations of R_p, $N_{p,f}$, N_m, and N_w with the sodium peroxodisulfate concentration ([SPS]):

$R_p \propto [SPS]^{0.54}$, $N_{p,f} \propto [SPS]^{-0.2}$, $N_m \propto [SPS]^{-0.47}$, and $N_w \propto [SPS]^{0.15}$.

These data indicate that the increase in $N_{p,f}$ with [SPS] is much more pronounced by homogeneous nucleation in comparison with monomer droplet nucleation. The dependencies of R_p and $N_{p,f}$ on [SPS] indicate that the increased R_p

with [SPS] should be attributed to the increased formation of polymer particles by homogeneous nucleation (no barrier to the entering radicals) and the extent of radical compartmentalization of radicals. The increase of N_w and decrease of $N_{p,f}$ and N_m with [SPS] result from the mixed modes of particle nucleation.

The compound DMA reduces R_p in the St mini-emulsion polymerization stabilized by NP40 and initiated by SPS or AIBN [17]. At low [DMA], the D vs. conversion data can be described by a curve with a minimum followed by a plateau. An opposite trend is observed for the run with higher [DMA], where D first increases to a maximum and then decreases to a plateau as the polymerization proceeds. This behavior can be attributed to increased accumulation of NP40 in the monomer phase (or decreased water solubility of monomer and NP40) with increasing DMA. Incorporation of DMA into the formulation greatly decreases the diffusion of NP40 through the aqueous phase. The authors interpreted these results in terms of several competitive events such as coalescence of the monomer droplets, nucleation in the monomer droplets and monomer-swollen micelles, formation of particle nuclei in the aqueous phase, and particle growth. Furthermore, the complex situation arises from the steric stabilization effect provided by NP40, which is greatly reduced at 80 °C and the CMC of the miniemulsion system is far below the NP40 concentration ([NP40]) used. The values of $N_{p,f}/N_{m,i}$ smaller than 1 were achieved in the runs with higher [DMA]. It was postulated that monomer droplet nucleation predominates in the course of polymerization. R_p, $N_{p,f}$ or $N_{p,f}/N_{m,i}$ increases with increasing [NP40]. The increase in R_p originates from micellar or homogeneous nucleation and increased stability of polymer particles as [NP40] is increased. During the early stage of polymerization, the degree of coalescence of the monomer droplets decreases rapidly with increasing [NP40]. The high oil-solubility of NP40 is expected to influence the colloidal parameters of the reaction system, that is, NP40 can also act as hydrophobe.

The concentrations of monomer droplets and latex particles are relatively insensitive to changes in [SPS]. On the contrary, R_p increases significantly with increasing [SPS]. The ratio $N_{p,f}/N_{m,i}$ increases with increasing [SPS] as a result of homogeneous nucleation. Similarly, the weight percentage of the extremely hydrophobic dye ultimately incorporated into the latex particles (P_{dye}) decreases significantly with increasing [SPS], which is attributed to the increased number of latex particles formed via micellar or homogeneous nucleation. However, the increased R_p with [SPS] is due to the increased number of latex particles produced by homogeneous nucleation which show a higher polymerization activity. A similar behavior concerning R_p and $N_{p,f}$ was also observed in mini-emulsion polymerization initiated by AIBN. A difference in the D vs. time data exists between the SPS and AIBN initiated polymerizations. The D vs. time dependence for the AIBN initiated polymerization is described by a curve with the initial strong increase in D followed by a plateau. Both $N_{p,f}/N_{m,i}$ and P_{dye} decrease with increasing AIBN concentration ([AIBN]). These data imply that the fraction of AIBN dissolved in water takes part in the initiation reaction and aqueous phase nucleation. Thus, the fraction of AIBN dissolved in water and the AIBN-derived primary radicals desorbed from the monomer droplets initiate the aqueous phase polymerization. In addition, the maximal P_{dye} (ca. 66.4%) indicates that only a fraction of the monomer droplets initially present in the mini-emulsion polymerization system can be successfully transformed into latex particles. The lower colloidal stability of the AIBN initiated mini-emulsion polymerization, as shown by the coagulum formation, results from the reduced stability of primary particles in the absence of SO_4^{2-} groups derived from SPS on the particle surface. This then promotes the particle agglomeration and leads to a value of $N_{p,f}/N_{m,i}$ less than 1.

In the St mini-emulsion polymerization stabilized by NP40, R_p decreases with increasing [DMA], and increases with increasing [NP40], [SPS] and [AIBN] [17]. The polymerization was carried out above the cloud point of NP40 and, therefore, the steric stabilization effect of NP40 is greatly reduced at 80 °C. For the AIBN containing reaction system, D (D_f=ca. 200 nm) increases to a plateau as polymerization proceeds, while D (D_f=ca. 250 nm) goes through a maximum after the start of polymerization for the SPS containing system. The subscript of D_f represents the value of D determined at the end of polymerization. A much larger amount of coagulum was collected in the AIBN-containing latex product than in the SPS counterpart. The amount of coagulum decreases with increasing [AIBN]. By contrast, the reverse is true for the SPS initiated mini-emulsion polymerization. The negative charges on the primary particle surface derived from SPS make the primary particles more stable in comparison to those produced in the AIBN initiated mini-emulsion polymerization. The increased water-phase termination of the SPS-derived radicals decreases the concentration of surface active oligomer and the participation of the oligomer in the homogeneous particle nucleation and stabilization. Monomer droplet nucleation becomes more important when [DMA] increases or [NP40], [SPS] and [AIBN] decreases. The ratio $N_{p,f}/N_{m,i}$ varies in the range of 0.3–0.8 and this implies the significance of coalescence of latex particles during polymerization. The release of non-micellar NP40 from the monomer droplets during polymerization can induce the secondary (homogeneous) nucleation.

It was reported that the Ostwald ripening effect is depressed when the nonionic emulsifier NP40 is incorporated into the stabilization system (SDS/DMA) [16]. When [NP40] increases, R_p and N_w decrease but N_m increases. This indicates that NP40 promotes the monomer droplet nucleation and the kinetics of mini-emulsion polymerization. Incorporation of NP40 into the emulsifier mixture leads to a significant deviation from the micellar nucleation model [113, 114]. This model is applicable only for the emulsion polymerization stabilized by SDS alone. The latex particles stabilized only by SDS or NP40 are quite stable judging from the level of coagulum produced in the course of polymerization. For the mixed emulsifier system, transition from a stable to an unstable colloidal system occurs above a critical [SPS]. The stability of primary particles formed by homogeneous nucleation decreases as a result of the addition of NP40. The decreased diffusion degradation of monomer droplets slows down the monomer transport through the aqueous phase as well as the formation of surface active oligomeric radicals. On the contrary, the increased termination of primary radical reduces the extent of participation of the surface active radicals in the nucleation and stabilization of primary particles. The product of the termination of primary radicals is the inactive water-soluble oligomer. Furthermore, the low level of free emulsifier caused by the increased total droplet surface area reduces the stabilization significantly and, therefore, increases the formation rate of coagulum. Indeed, the CMC of the aqueous SDS solution is greatly reduced by the incorporation of NP40 into the stabilization system [115]. Furthermore, a perceptible minimum appears in every curve of the surface tension vs. the logarithm of total emulsifier concentration over all the mixed emulsifier compositions studied at 25 °C. On the other hand, none of the surface tension curves shows a detectable minimum at 80 °C. This complex behavior can be interpreted in terms of the increased solubility of NP40 in the monomer phase with increasing temperature. The resultant latex particles are relatively monodisperse and the average particle diameter is varied in the range of 50–60 nm with increasing levels of NP40 in the mixed emulsifier SDS/NP40 mixture. The increase in the level of the non-ionic emulsifier is expected to increase the latex particle size, as demonstrated by the emulsion polymerization of alkyl acrylates stabilized by the anionic/non-ionic emulsifier mixture [112]. However, the reverse is true; at ca. 30% conversion the particle size decreases with increasing levels of NP40 in the emulsifier feed. This cannot be attributed to the steric stabilization mechanism, but rather to the increased stability of the monomer droplets. Furthermore, the milky monomer emulsion is converted by polymerization to a fine translucent polymer latex. Accumulation of hydrophobe (St oligomer formed by the chain transfer to NP40) might increase the stability of monomer droplets and the length of particle nucleation (monomer droplet nucleation). The parameter y' appearing in the relationship $N_p \propto [\text{emulsifier}]^{y'}$ increases from 0.6 to 1.8 with increasing [NP40]. This is most likely due to the relatively oil-soluble non-ionic emulsifier. Consequently, the amount of emulsifier available for the formation of micelles or stabilization of polymer particles is smaller than that initially charged in. Thus, above the critical concentration of NP40 at which the monomer phase is saturated with NP40, the non-ionic emulsifier starts to form micelles and subsequently stabilize the polymer particles nucleated in emulsion polymerization. The parameter x' ($N_p \propto [\text{initiator}]^{x'}$) first increases to a maximum and then decreases with an increase in [NP40]. The reaction order x' is a complex function of the concentration of free non-ionic emulsifier, the chain transfer (termination) to non-ionic emulsifier and particle nucleation. Therefore, this relationship is rather empirical.

Monomer droplet nucleation plays an important role in both the DMA (1) and SMA (2) containing mini-emulsion polymerization of St initiated by AIBN [116]. On the other hand, increasing [SPS] promotes the homogeneous nucleation in the St mini-emulsion polymerization system (see above). For example, for the SMA series, the degree of homogeneous nucleation decreases in the series:

SPS>MIX>AIBN

where MIX represents a mixture of SPS and AIBN used to initiate the St mini-emulsion polymerization. For the copolymerization of DMA/St initiated by AIBN, the kinetic data suggest that segregation of free radicals among the latex particles predominates in the polymerization with the initial monomer droplet having a diameter $(D_{m,i})<300$ nm. On the other hand, the pseudo-bulk polymerization kinetics operate when $D_{m,i}>300$ nm. As $D_{m,i}$ is increased, the number of particle nuclei generated by homogeneous nucleation becomes smaller, whereas monomer droplet nucleation becomes more important. This is attributed to the cage effect associated with the initiator radicals produced in pairs inside the large latex particles.

For the DMA (1) or SMA (2) containing mini-emulsion polymerization of St, the following relationships hold:

$$R_p \propto [AIBN_1]^{1.07} \text{ and } R_p \propto [AIBN_2]^{0.55}$$

$$N_{p,f} \propto [AIBN_1]^{0.67} \text{ and } N_{p,f} \propto [AIBN_2]^{0.6} \text{ (up to 11.2 mM)}$$

$$N_m \propto [AIBN_1]^{0.43} \text{ and } N_m \propto [AIBN_2]^{0.5} \text{ (up to 8.5 mM)}$$

$$N_w \propto [AIBN_1]^{4.7} \text{ and } N_w \propto [AIBN_2]^{1.3} \text{ (up to 11.2 mM)}.$$

These results show that R_p increases with increasing [AIBN] and the degree of increase is more pronounced in the runs with DMA. For the copolymerization of DMA/St, the polymerization rate per particle $(R_p/N_{p,f} \propto [M]_p \, \bar{n})$ increases with increasing [AIBN]. On the other hand, $R_p/N_{p,f}$ is relatively insensitive to changes in [AIBN] for the SMA/St counterpart. For both the DMA and SMA series, the particle formation rate is suppressed at high [AIBN]. The dependence of $N_{p,f}$ on [AIBN] is described by a curve with a maximum at a certain level of AIBN. $N_{p,f}$, N_m and N_w decrease significantly above a critical [AIBN] (8.5 or 11.2 mM). This may be attributed to the increased termination of primary radicals, which decreases the concentration of the initiating radicals. The reaction order 1.07 in the relationship $R_p \propto [AIBN_1]^{1.07}$ is connected with the high homogeneous nucleation activity of the DMA containing mini-emulsion polymerization. On the contrary, monomer droplet nucleation leads to the lower reaction order of 0.55 in the relationship $R_p \propto [AIBN_2]^{0.55}$ for the SMA counterpart.

For the DMA (1) containing mini-emulsion polymerization of St initiated by AIBN, the following relationships hold:

$$R_p \propto [SDS_1]^{0.56}, N_{p,f} \propto [SDS_1]^{0.69}, N_m \propto [SDS_1]^{0.21} \text{ and } N_w \propto [SDS_1]^{1.64}.$$

The reaction order 0.56 obtained from $R_p \propto [SDS_1]^{0.56}$ indicates the emulsifier-flooded condition (the monomer droplet surface not saturated with SDS). However, the increased coverage of the droplet surface by emulsifier is accompanied by the enhanced homogeneous nucleation, i.e., N_w increases significantly with increasing $[SDS_1]$. Monomer droplet nucleation predominates in the particle formation process for the run with the lowest $[SDS_1]$ (4 mM). By contrast, mixed modes of particle nucleation are operative in the polymerizations with

$[SDS_1] \geq 3 \cdot 10$ mM. These data also can be used to explain the decreased $R_p/N_{p,f}$ with $[SDS_1]$. As more latex particles are generated by homogeneous nucleation, both the cage effect associated with the initiator radicals produced in pairs inside the particles and segregation of the absorbed radicals among the discrete particles will become more pronounced. As a result, the number of radicals which can be accommodated into the particles will decrease with increasing $[SDS_1]$.

The foregoing results as well as those summarized in Tables 1 and 2 indicate that the exponent x' obtained from the relationship $N_p \propto [I]^{x'}$ can be taken as a measure of the particle nucleation and it is proportional to the ratio of the final number of polymer particles generated to the initial number of mini-emulsion droplets ($N_{p,f}/N_{m,i}$). If the particle nucleation process is primary controlled by the radical entry into monomer droplets (monomer droplet nucleation), the value of $N_{p,f}/N_{m,i}$ should be around unity and the value of x' should be ≤ 0.1 [12]. A value of $N_{p,f}/N_{m,i}$ lower than 1 implies incomplete monomer droplet nucleation or coalescence among the interactive droplets. The value of x' greater than 0.1 may indicate the influence of micellar nucleation, since one droplet feeds monomer to more than one micelle in the classical emulsion polymerization. Furthermore, the compartmentalization of radicals is increased. Table 1 indicates that the values of x' for the mini-emulsion polymerization of hydrophobic St with CA as the coemulsifier are relatively large and much larger than those for the mini-emulsion polymerization of polar MMA in the presence of LPO or DDM. On the contrary, the addition of unsaturated polar SMA strongly decreases the value of x' for the St mini-emulsion polymerization. The same behavior was observed for the MMA mini-emulsion polymerization with the unsaturated VH. Thus, the chemically bound polar hydrophobe in the particle surface layer significantly promotes monomer droplet nucleation. The low values of x' were also achieved for the MMA mini-emulsion polymerization with HD, LPO and DDM. The latter behavior can result from the synergistic effect of PMMA formed *in situ* as a hydrophobe. On the contrary, the relatively high hydrophilicity of MMA favors homogeneous nucleation and, thereby, increases x'. Indeed, the presence of the hydrophilic co-monomer VAc in the BA mini-emulsion polymerization enhances homogeneous nucleation and leads to the strong rise in x'. In addition, MMA may act as a cosolvent (coemulsifier), which promotes the formation of a fine monomer dispersion [117].

The DMA or SMA containing mini-emulsion polymerization can be discussed in terms of its coemulsifier (hydrophobe) efficiency. The value of z ($N_p \propto [CoE]^z$) for the MMA and St mini-emulsion polymerizations increases in the series (Table 1):

– 0.35 (DMA, St)<0.11 (LPO, MMA)<0.22 (HD, MMA)<0.36 (DDM, MMA).

N_p decreases with increasing [CoE] for the DMA containing mini-emulsion polymerization of St. A similar effect was observed for the SMA containing mini-emulsion polymerization of St initiated by SPS for the dependence of N_p on [initiator]. This was attributed to the variation of homogeneous nucleation or mixed modes of particle nucleation. This can also be connected with the increased stability of monomer droplets with [DMA] and droplet nucleation (decreased homogeneous nucleation). Incorporation of DMA into the formulation greatly decreases the diffusion of monomer (reactants) through the aqueous phase. Furthermore, the shift in the copolymer composition with increasing [DMA] (or [SMA]) varies the interaction of SDS emulsifier with the polymer particle surface and, thereby, the stability of latex particles. For the mini-emulsion polymerization with LPO, HD or DDM, N_p increases with increasing [CoE]. Furthermore, the increase of N_p is the most pronounced for the polymerization with DDM (DDM, as a polar hydrophobe strongly increases the particle population via monomer droplet nucleation). The saturation of droplet surface with coemulsifier species increases the stability of monomer droplets and their surface area. Besides, DDM also acts as a short-stopper of the polymer growth events. Under these circumstances, the accumulation of low molecular weight byproducts in the monomer droplets increases the stability of the monomer droplets. In the case with LPO, the accumulation of LPO and its radical fragments also increases the stability of the oil-water interfacial layer, but the termination of primary radicals might reduce the initiator efficiency (or particle formation). The termination of primary radicals is not expected for the mini-emulsion polymerization with HD and therefore, the particle nucleation rate is increased.

The activity of several unsaturated coemulsifiers [alkyl methacrylates (RMA) such as DMA, hexyl methacrylate (HMA), and bornyl methacrylate (BMA)] was evaluated in the emulsion polymerization systems stabilized by SDS or a nonionic emulsifier of the polyoxyethylene type [118]. The length of the particle nucleation period is very long and it increases with increasing alkyl chain length of RMA. This is attributed to the interfacial barrier for radical entry into monomer droplets and the suppressed water-phase polymerization that generates surface-active oligomeric radicals. The close-packing of emulsifier and coemulsifier (RMA) at the mini-emulsion droplet surface provides a barrier to the entering radicals. The influence of the close-packed structure is much more significant in runs with SDS as compared to the non-ionic emulsifier counterpart. This results from the predominant location of SDS in the oil-water interfacial zone while the non-ionic emulsifier counterpart can penetrate into the monomer droplet core. The length of the particle nucleation period is greatly prolonged in the copolymerization of RMA with trimethylolpropane trimethacrylate (TMPTMA) and it increases with increasing fraction of TMPTMA in the feed. The addition of TMPTMA as both hydrophobe and crosslinker clearly extends the particle nucleation period. This is due to the lower monomer concentration in water and water-phase polymerization, the slower transfer of monomer from droplets to the reaction loci and the increased importance monomer droplet nucleation as a result of the increased stability of monomer droplets provided by TMPTMA. Besides, the reduced particle growth caused by the relatively low reactivity of the bulky TMPTMA promotes the accumulation of a small amount of hydrophobic polymer in the monomer droplets.

A common feature of the emulsion polymerization of allyl methacrylate (AMA) is the slow rise to a maximal R_p and two rate regions [119]. The increase of R_p is ascribed to the increased number of latex particles, while the decrease of R_p is attributed to the decreased monomer concentration at the reaction loci. The slow accumulation of polymer in the monomer droplets (due to the very low reactivity of AMA) retards the monomer droplet degradation and the transfer of monomer from monomer droplets to the reaction loci but enhances the monomer droplet nucleation. R_p/N_p increases with increasing particle size. The slight increase in R_p with increasing [SDS] is due to the high coverage of the droplet surface by SDS.

3.5
Other Kinetic Aspects

The mini-emulsion polymerization technique offers a better control over the copolymer composition because incorporation of the constituent monomers into the emulsion polymer product is not governed by the mass transfer process or its water solubility [37, 74, 111]. The resultant bulk or solution polymer composition obeys the classical copolymer equation and it is governed by the reactivity ratios of the constituent monomers. In contrast, the copolymer composition of the polymer product prepared by emulsion polymerization is controlled by mass transfer, which may limit the incorporation of water-insoluble monomer into latex particles. The more water-soluble monomer polymerizes preferentially because it can diffuse through the aqueous phase to the reaction loci, followed by the less water-soluble monomer at higher conversion. For example, in the mini-emulsion copolymerization of MMA and α-methylstyrene (MSt), the copolymer product comprises more MMA than the copolymer equation would predict. The mini-emulsion droplets would have a higher concentration of the less water-soluble co-monomer throughout the polymerization. Therefore, the droplet nucleation should yield copolymer rich in the more hydrophobic comonomer. On the contrary, homogeneous nucleation should lead to copolymer with a higher MMA content at the beginning of the polymerization because MMA can cross the continuous aqueous phase more readily. A similar behavior was observed in the emulsion copolymerization of acrylonitrile (AN) and butyl acrylate (BA) [38]. The initial copolymer composition is more abundant in the hydrophilic AN. The evolution of the BA/VAc copolymer composition, however, shows that the copolymer produced in mini-emulsion polymerization contains fewer VAc units than that produced in the conventional emulsion polymerization. The dynamic mechanical spectroscopy shows two distinct glass transition temperatures (T_g) for the copolymer product obtained from both polymerization systems, one for the BA-rich domain and the other for the VAc-rich domain. This is the result of the shift in the copolymer composition with the progress of polymerization. The copolymer formed in mini-emulsion polymerization exhibits a lower T_g for the BA-rich domain than that formed in the conventional emulsion polymerization. Furthermore, the shapes of the storage and loss moduli (E′, E″) data indicate less mixing between the two phases (BA-rich domain and VAc-rich domain) in the latex particles produced in mini-emulsion copolymerization.

The low- or medium-molecular weight hydrophobe (coemulsifier) affects the distribution of monomer between different phases of the heterogeneous polymerization system. The presence of hydrophobe in the bulk monomer does not greatly influence the polymerization kinetics unless a mini-emulsification process is used. In this case, the coemulsifier species effectively stabilize the mini-emulsion droplets, which then become the main locus of particle formation. This will result in a drastic change in the polymerization mechanism. The capability of retaining monomer in the droplets is greatly enhanced when coemulsifier is added to the monomer phase. In addition, the point at which the droplets start to disappear is shifted to higher conversion as the coemulsifier concentration is increased. The importance of the aqueous phase or oil phase polymerization was further investigated by the San Sebastian group [120]. In St emulsion and mini-emulsion polymerizations initiated by AIBN, $R_p/N_{p,f}$ increases when the weight ratio of monomer to water increases. Asua et al., interpreted these data in terms of desorption of initiator radicals from latex particles to the aqueous phase. Furthermore, $R_p/N_{p,f}$ increases with increasing total monomer droplet surface area. $R_p/N_{p,f}$ obtained from the mini-emulsion polymerization is much lower by 1–2 orders of magnitude than that obtained by the classical emulsion polymerization. Incorporation of HD into the formulation greatly decreases the diffusion of AIBN through the aqueous phase. The presence of seed latex particles and growth of these particles indicate that, besides the molecular diffusion of AIBN through the aqueous phase, other mechanism(s) can contribute to the whole transport process. Transport of the AIBN single-radicals from the monomer droplets into seed particles via droplet/particle collisions probably occurs. Rearrangement of the reaction ingredients during the droplet/particle collisions and the decrease in the oil-water interfacial tension promotes the escape of the AIBN radicals out of the cage volume. The reversible collision and separation generates two single-radicals in two interactive particles, which then initiates or terminates the polymer chain growth process. The same behavior can be expected for the irreversible particle collisions. Here, rearrangement of the reagents promotes the separation of radicals from the cage effect.

The basic principle behind the preparation of a stable mini-emulsion lies in the decrease of solubility of monomer in the aqueous phase. This approach was applied to the fine emulsion polymerization and copolymerization of MMA, EA, nonyl methacrylate (NMA) and methacryloyl-terminated polyoxyethylene macromonomer (PEO-MA) initiated by UV light at room temperature by Capek [121, 122]. At low temperature, the monomer droplet degradation is suppressed due to the lower water-solubility of monomer. Under these circumstances, the dependence of the rate of polymerization of MMA, EA or NMA is described by a curve with two maximal rates or four distinct non-stationary rate intervals, typical for the mini-emulsion polymerization. The effect of the restricted monomer droplets degradation is also considered as the result of accumulation of hydrophobic monomer (NMA) or oligomer (polymer) in the microdroplets. The specific interaction between the coemulsifier (monomer) and emulsifier is also operative in this polymerization system. This results from the high concentration of electrolyte (emulsifier). Under the circumstances, the dissociation of emulsifier is depressed but the close-packing of emulsifier/coemulsifier via both hydration and hydrophobic interactions is favored. The first maximum on the R_p-conversion curve is a basic characteristic of the micro-emulsion polymerization system [117]. This is due to the increasing particle concentration and decreasing monomer concentration at the reaction loci with increasing conversion. The second maximal R_p can result from the longer preservation of monomer droplets (highly monomer-swollen polymer particles) and accumulation of polymer and radicals within these monomer/polymer particles. Accumulation of polymer transforms the thermodynamically stable microdroplets to stable minidroplets (the increased turbidity of reaction system). The addition of PEO-MA decreases R_p and polymer molecular weight but increases the average latex particle size. The graft oligomer or copolymer (PMMA-graft-PEO, PEA-graft-PEO or PNMA-graft-PEO) is incorporated into the particle shell and, thereby, transforms the electrostatically-stabilized polymer particles to the electrosterically-stabilized polymer particles. However, the polymer latex product is very fine and this can be attributed to the incorporation of graft copolymer (acting as coemulsifier) into the particle shell, inducing interaction between emulsifier and coemulsifier (graft copolymer).

A significant deviation from the micellar nucleation model was observed in the St emulsion polymerization stabilized by SDS (above CMC) and initiated by SPS at different temperatures using a long pre-emulsifier period (the monomer emulsion was mixed at 400 rpm for 2 h at 60 °C) [123, 124]. The R_p vs. conversion curves show two distinct non-stationary rate regions and a shoulder occurring at high conversion, while the stationary rate region is quite short or does not appear at all. R_p and $N_{p,f}$ are proportional to $R_i^{0.27}$. A significant population of tiny monomer droplets still remains in the reaction mixture up to ca. 90% conversion. The overall activation energy is quite close to the one typical for emulsion polymerization, but it increases with increasing conversion presumably due to the increased contribution of polymerization within the monomer droplets. The shoulder appearing on the R_p vs. conversion curve, the appearance of a maximum at high conversion on the $(\bar{n})$ vs. conversion curve, the increase of activation energy with increasing conversion, the low particle nucleation activity, the long particle nucleation period, the relatively narrow PSD of latex particles, and independence of PSD on conversion deviated from the micellar nucleation model. Furthermore, the reaction order 0.27 from the relationship $R_p \propto R_i^{0.27}$ deviates from the micellar nucleation model. Deviation of the St emulsion polymerization from the micellar nucleation model is induced by the accumulation of hydrophobic oligomer, byproducts or polymer in the monomer droplets, which increases the stability of monomer droplets as well as the length of the particle nucleation period. Under these circumstances, the apparent monomer-starved condition (the monomer droplets are still present) significantly influences the emulsion polymerization. The high ratio of the absorption rate of radicals by particles to the formation rate of radicals in water (much above 1) supports the additional formation of radicals in the monomer droplets by the thermally induced initiation. This ratio is also a function of the uncharged radical concentration in the reaction system. The important role of uncharged styrenic radicals might be due to the relatively low activation energy per particle (ca. 34 kJ·mol⁻¹, the low energetic barrier for the entering radicals). The desorption of uncharged radicals becomes more important and results in lower values of $\bar{n}$ ($\bar{n}$ <0.5) as temperature is raised above 60 °C.

The latex particles of diameter ca. 50–80 nm were also prepared by the St emulsion polymerization initiated by the redox system of APS/sodium thiosulfite and stabilized by a non-ionic emulsifier (Tween 20, polyoxyethylene sorbitan monopalmitate) [125]. The R_p data as a function of conversion can be described by a curve with a maximum and region II cannot be identified. The maximal R_p is proportional to the –0.45 and 1.5 power, N_p to the 0.32 and 1.3 power, and the polymer molecular weight to the –0.62 and –0.97 power of the initiator and emulsifier concentrations, respectively. A significant reduction in the turbidity at ca. 20% conversion was observed when the emulsion turned into a translucent latex. Deviation from the micellar nucleation model is attributed to the solubility of the non-ionic emulsifier in monomer, the high level of non-micellar aggregates, the rather thick oil-water interfacial layer, and the transformation of emulsion to mini-emulsion. The rapid decrease of the polymer molecular weight with increasing emulsifier concentration is attributed to the pseudo-bulk kinetics and chain transfer reaction promoted by the high level of non-ionic emulsifier at the reaction loci.

The coarse monomer emulsion was transformed to the fine polymer latex during polymerization [126]. For emulsion polymerization of St stabilized by Tween (Tw), the shape of the R_p vs. conversion curve is a function of the reaction temperature. In runs A (50 °C) and D (80 °C), the R_p vs. conversion curve shows two distinct non-stationary rate intervals. In the runs B (60 °C) and C (70 °C), the two distinct non-stationary rate intervals and two maximal rates appear in the course of polymerization. In the run D (with the lowest amount of Tw in the aqueous phase), coagulative nucleation is operative in the polymerization system. The conversion of the coarse emulsion to the fine emulsion results from the particle nucleation up to high conversion and the depressed particle growth. The continuous particle nucleation is caused by release of non-micellar emulsifier to the aqueous phase, thereby leading to the increased emulsifier concentration for particle nucleation and stabilization. The depressed particle growth results from the chain transfer to emulsifier and dilution of monomer by non-ionic emulsifier. The relatively low apparent value of E_o=46.1 kJ·mol⁻¹ is interpreted in terms of the ratio of the steric to electrostatic stabilization. The continuous change of the stabilization mechanism from the electrostatic to electrosteric type with increasing temperature then decreases E_o. The transformation of the stabilization mechanism is the reason why the dependency of R_p or N_p on R_i (0.22 or 0.67 power) deviates from the micellar nucleation model. Increased temperature and entry of charged oligomer into particles assist in developing the electrostatic type of interface (with a low barrier or no barrier to the entering radicals). The hairy-like interface comprising non-ionic emulsifier forms a strong barrier to entering radicals.

The deviation from the micellar nucleation model observed in the St emulsion polymerization stabilized by anionic or non-ionic emulsifier is attributed to the increased contribution of monomer droplet nucleation (or mixed modes of particle nucleation). That is, monomer droplets do not act only as a reservoir of monomer. For example, a significant population of monomer droplets at high conversion and the very low mass ratio of St/PSt (0.15) in the monomer-swollen latex particles were found in the St emulsion polymerization system [123]. The video-enhanced microscopy measurement proved the presence of the water-in-oil-in-water (w/o/w) droplets in the St monomer droplets and a significant population of monomer droplets still remained in the reaction mixture beyond region II [39]. Furthermore, it was estimated that ca. 5% of SDS molecules go into the monomer phase and form reverse micelles or tiny w/o/w droplets inside the St phase (i.e., monomer droplets). The ultrasonification process is connected with the rapidly increased oil-water interfacial area as well as the significant reorganization of the droplet clusters or droplet surface layer. This may lead to the formation of additional water-oil interfaces (inverse micelles) and, thereby, decrease the amount of free emulsifier in the reaction medium. The formation of double droplets with the one-step process is promoted by ultrasonification, addition of different organic compounds (polymer, oligomer, etc.), and increasing the viscosity gradient between the dispersed phases [127]. In addition, these double emulsion droplet structures are more favored in the presence of mixed non-ionic/anionic emulsifiers. Thus, the addition of hydrophobe also can be related to the increased fraction of double droplets. Under these circumstances, the free emulsifier concentration is rapidly decreased as well as the monomer transfer rate. Furthermore, the inverse micelles in the double emulsion are directly connected with the transport phenomena of all reactants involved. This might be the case for the mini-emulsion polymerization system. However, this speculation is open to experimental verification.

The hairy particles stabilized by non-ionic emulsifier (electrosteric or steric stabilization) enhance the barrier for entering radicals and differ from the polymer particles stabilized by ionic emulsifier [35]. For example, the polymer lattices with the hairy interface have much smaller values of both the radical entry (ρ) and exit (k_{des}) rate coefficients as compared to the thin particle surface layer of the same size [128,129]. The decrease of ρ in the electrosterically stabilized lattices is ascribed to a "hairy" layer which reduces the diffusion of oligomeric radicals, so that these radicals may be terminated prior to actual entry. For the electrostatically stabilized lattices with the thin interfacial layer, exit of radicals occurs by the chain transfer reaction [35]. This chain transfer reaction results in a monomeric radical which then exits out of the particle by diffusing through the aqueous phase and this event is competing with the propagation reaction in the particle [130]. The decrease of k_{des} in the electrosterically stabilized latex particles with a rather thick interfacial layer was interpreted by assuming that the diffusion of radicals in the hairy layer is slow. The formation of a close-packed structure at the droplet-water interface via the interaction between the emulsifier and coemulsifier is expected to influence both the entry and exit of radicals. The entry/exit ratio depends on the nature of the particle (droplet)-water interface and the radicals produced in both the aqueous phase (surface active oligomeric radicals) and monomer droplets (transferred monomeric and emulsifier radicals). The charged oligomeric radicals formed in the aqueous phase and much smaller transferred monomeric radicals without charge formed in the monomer phase indicate the different radical entry efficiency.

An inhibitor (radical scavenger) influences the kinetic and colloidal parameters in a very complex way. It affects the rate of emulsion polymerization and the colloidal parameters of the emulsion polymer [36]. From the classical point of view, an inhibitor deactivates the reaction loci and the polymerization should proceed after the consumption of inhibitor. Table 2, however, shows that the addition of inhibitor decreases the sensitivity of both the classical emulsion and mini-emulsion polymerizations to the initiator concentration. This can be attributed to the short-stopping of the propagation events and the increased monomer droplet nucleation. Furthermore, the PMMA-stabilized mini-emulsion polymerization kinetics is less sensitive to the presence of inhibitor than the conventional emulsion polymerization kinetics. Besides, the ratio $N_{p,f}/N_{m,i}$ is very close to 1 for the PMMA-stabilized mini-emulsion polymerization with and without inhibitor. The contribution of inhibitor to the stability of minidroplets can results from the accumulation of hydrophobic oligomer (byproduct) during the inhibition period. The retardation of polymerization within the monomer droplets by the immobilized inhibitor results in the accumulation of an oligomer among a large number of monomer droplets and, consequently, the stabilization of monomer droplets against the diffusional degradation. Furthermore, the decreased growth of latex particles during polymerization by the transfer of oil-soluble inhibitor from the monomer droplets to reaction loci increases the probability for the entry of radicals to a larger population of monomer droplets.

Controlled free-radical polymerization of styrene was performed in a mini-emulsion medium using a degenerative transfer process with iodine atom exchange [131]. Its efficiency was shown to be low (50%) in conventional batch emulsion polymerization since the polymer had a molar mass higher than expected. This was explained by a low rate of diffusion of the perfluorinated transfer agent through the water phase, from the monomer droplets to the active latex particles. This means that the transfer agent was not totally consumed before complete monomer conversion, that is, the hydrophobic perfluorohexyl iodide $(C_6F_{13}I)$ accumulates during polymerization in the non-nucleated monomer droplets or SDS micelles. This can be a reason why the rate of polymerization and the particle diameter were not affected by the increased amount of the per-fluorinated compound. However, when using a batch mini-emulsion process, 100% efficiency was reached, and the experimental molar masses fit well with the theoretical ones. With the mini-emulsion technique, the transfer agent (acts as hydrophobe) could be directly located in the polymerization locus without transportation limits. Fast and complete polymerizations were obtained when using the water-soluble (ACPA) initiator, similar to the case for the previous emulsion polymerizations. The use of the oil-soluble AIBN as a radical initiator led to a comparatively slower polymerization. The mini-emulsion polymerization, however, was faster than the emulsion polymerization. An increase of the molar mass with monomer conversion (up to $M_n=1.6 \times 10^4$ for emulsion and 5×10^4 for mini-emulsion) was observed together with a decrease of M_w/M_n from 2 to 1.5–1.6 (mini-emulsion). This result was attributed to the relatively low transfer constant of the transfer agent. The proportionality of M_n with monomer conversion was observed when styrene was continuously fed into the polymerization medium. In that case, however, the molar mass distribution broadened with monomer conversion (from 1.3 to 2.3), and this was explained by a decrease in the rate constant of the activation/deactivation reaction with chain length and viscosity of the reaction medium.

Langfester et al. [132] have shown that the principle of aqueous mini-emulsions can be transferred to non-aqueous media. In direct mini-emulsions using polar media such as formamide or glycol instead of water and hydrophobic monomer, the mini-emulsion stability was, as in the case of the aqueous systems, obtained by the hydrophobe which prevents monomer droplet degradation. In the case of inverse systems, hydrophilic monomers were mini-emulsified in a non-polar media (cyclohexane, hexadecane, etc.). To provide osmotic stabilized droplets, simply water or salt was added as lipophobe to the monomer phase. It was shown that such direct and inverse mini-emulsions showed long-term stability against Ostwald ripening. The polymer particles had a diameter of ca. 70 nm in the direct mini-emulsion. In the inverse micro-emulsion, the polymer particles in the range from 50 to 200 nm were prepared. Surface tension measurements showed incomplete coverage of the polymer particles with emulsifier in both systems. Steric stabilizers (non-ionic) based on poly(ethylene oxide) tails are by far more efficient than ionic stabilizers, which is attributed to a low degree of ion solvation and degree of dissociation in formamide.

4
Conclusion

The mini-emulsion polymerization process can be divided into four major regions (Scheme 2). In the first region, significant nucleation of polymer particles takes place and, as a result, the maximal polymerization rate is achieved. Immediately after the start of the second region, the polymerization rate (R_p) begins to fall off. The decrease in R_p is attributed to the reduction of the monomer concentration in the monomer-swollen polymer particles as the polymerization proceeds. The decreased monomer concentration in the reaction loci becomes more pronounced when the number of latex particles generated per unit volume of water (N_p) is increased and the diffusional degradation of monomer droplets is greatly depressed. The reduced diffusional degradation is caused by the accumulation of polymer in the nucleated monomer droplets. The second maximal R_p observed in the third region is due to the gel effect. In the fourth region, R_p continues to decrease simply due to the depletion of monomer in the polymerizing system.

One of the key features of this reaction mechanism is the particle nucleation beyond the first maximal R_p. Furthermore, the disappearance of monomer droplets in the conventional emulsion polymerization at ca. 30–50% conversion results from the transfer of monomer from the monomer droplets to the locus of polymerization (polymer particles). In mini-emulsion polymerization, the concentration of monomer decreases and monomer droplets may exist throughout the polymerization. In the context of the micellar nucleation model, both N_p and the concentration of monomer in the particles in the constant reaction rate region contribute to R_p. Therefore, the first maximal R_p observed in the course of mini-emulsion polymerization does not necessarily correspond to the end of particle nucleation. This is because N_p may increase in the course of polymerization, but the contribution of the increased N_p to R_p can be outweighed by the decreased monomer concentration in the reaction loci.

The presence of coemulsifier (hydrophobe) in the minidroplets, which are present in the colloidal system toward the end of polymerization, greatly reduces the free energy of mixing of monomer in the minidroplets. As a the result, the flux of monomer from the monomer droplets to the growing polymer particles is greatly reduced during the mini-emulsion polymerization. However, the hydrophobe cannot be transported from the monomer droplets to the polymer particles because of its extremely low water solubility. Thus, the concentration of coemulsifier in the monomer droplets is higher than that in the polymer particles. In this manner, monomer is retained in the monomer droplets to minimize the hydrophobe concentration gradient. The hydrophobe plays four major roles:

(1) A very large number of submicron monomer droplets is formed during the homogenization process and most of the emulsifier is adsorbed on the droplet surface. This is one of possible explanations for the low capture efficiency due to which the relative displacement of emulsifier molecules by the incoming oligomeric radicals is depressed.

(2) Upon initiation of the polymer reaction, these monomer droplets become the main locus of particle generation. At this stage of polymerization, the function of hydrophobe is to retain the monomer originally present in the nucleated droplets.

(3) As the polymerization proceeds, the presence of hydrophobe in the non-initiated monomer droplets reduces the concentration of monomer in the polymer particles.

(4) The presence of hydrophobe in the resultant polymer particles increases their swelling capacity.

The coemulsifier (hydrophobe) may play two additional roles during the mini-emulsion polymerization. It may hinder or retard the rate of interparticle monomer transfer by two different mechanisms:

(1) formation of an oil-water interfacial complex between the emulsifier and co-emulsifier, which may increase the interfacial resistivity of monomer droplets to entry and exit of the reacting species and

(2) retardation of monomer diffusion due to the osmotic pressure effect provided by the extremely low water solubility of hydrophobe.

The surface resistivity to the radical entry might also lead to the partial diffusional degradation of monomer droplets (the reservoir of monomer) and, therefore, the droplet size decreases with increasing conversion. The decrease of the droplet size increases the density of the interfacial layer formed by emulsifier and coemulsifier. This factor enhances the interfacial barrier of monomer droplets to the radical entry.

The kinetics of mini-emulsion polymerization using an oil-soluble initiator is less sensitive to changes in the level of initiator than that using a water-soluble initiator. In addition, the R_p data for the mini-emulsion polymerizations with an oil-soluble initiator are several times smaller than those with a water-soluble initiator. These findings were interpreted in terms of the higher initiation efficiency (increased homogeneous nucleation) when a water-soluble initiator was employed. The very low water-solubility of the oil-soluble initiator and the relatively fast R_p for the mini-emulsion polymerization support the idea that the single primary radicals are produced in the oil phase, which is probably caused by desorption of one of the initiator radicals from the cage and the radical hopping event during the interparticle collision. The same dependencies of R_p and N_p on the initiator concentration ([I]) imply that R_p is directly proportional to N_p. The degree of sensitivity of R_p and N_p is much less with the surface-active initiator (LPO). The opposite trend was observed in the system with unsaturated hydrophobe (SMA and DMA). Furthermore, the sensitivity of N_p to changes in [initiator]$^{x'}$ is reduced in the presence of a small amount of polymer.

The mini-emulsion polymerization kinetics is less sensitive to changes in the emulsifier concentration ([E]) than the conventional emulsion polymerization. In mini-emulsion polymerization, emulsifier is strongly adsorbed on the minidroplet surface and the fraction of free emulsifier micelles is lowered. Furthermore, the hydrophobic and hydration interactions between the emulsifier and coemulsifier (hydrophobe) increase the density of the droplet surface layer. The deviation from the mini-emulsion polymerization kinetics with conventional coemulsifier, however, can be observed in the reaction system with a chain transfer agent (DDM). N_p is very sensitive to changes in [E] when the amount of emulsifier used in the polymerization is relatively low. As soon as the monomer droplet surface is saturated with emulsifier, N_p is nearly independent of [E]. Most of the values of y' ($N_p \propto [E]^{y'}$) are relatively large, indicating that the monomer droplet surface is not saturated with emulsifier (Table 1). High [E] promotes the contribution of homogeneous nucleation or mixed modes of particle nucleation in the particle formation process. This is expected to be more pronounced for the hydrophilic monomers. However, the close-packed interfacial structure on the droplet surface can be reinforced by increasing [E] due to decreased degree of dissociation of emulsifier.

The ultrasonification process is connected with the rapidly increased oil-water interfacial area as well as the significant re-organization of the droplet clusters or droplet surface layer. This may lead to the formation of additional water-oil interface (inverse micelles) and, thereby, decrease the amount of free emulsifier in the reaction medium. This is supposed to be more pronounced in the systems with non-ionic emulsifier. Furthermore, the high-oil solubility of non-ionic emulsifier and the continuous release of non-micellar emulsifier during polymerization influence the particle nucleation and polymerization kinetics by a complex way. For example, the hairy particles stabilized by non-ionic emulsifier (electrosteric or steric stabilization) enhance the barrier for entering radicals and differ from the polymer particles stabilized by ionic emulsifier. The hydrophobic non-ionic emulsifier (at high temperature) can act as hydrophobe.

Formation of latex particles can proceed via the micellar nucleation, homogeneous nucleation and monomer droplet nucleation. The contribution of each particle nucleation mechanism to the whole particle formation process is a complex function of the reaction conditions and the type of reactants. There are various direct and indirect approaches to determine the particle nucleation mechanism involved. These include the variations of the kinetic, colloidal and molecular weight parameters with the concentration and type of initiator and emulsifier. There are some other approaches, such as the dye method where the latex particles generated via homogeneous nucleation do not contribute to the amount of dye detected in the latex particles since diffusion of the extremely hydrophobic dye molecules from the monomer droplets to the latex particles generated in water is prohibited. On the contrary, nucleation of the dye containing monomer droplets leads to the direct incorporation of dye into the polymer product. However, the dye also act as a hydrophobe and enhances the stability of monomer droplets as well as the monomer droplet nucleation.

Aknowledgements. The financial support by National Science Council, Taiwan is gratefully acknowledged. This research was also supported by the Slovak Grant Agency (VEGA) through the grants number 2/5005/98.

References

1. Guo JS, Sudol ED, Vanderhoff JW, El-Aasser MS (1992) J Polym Sci Polym Chem Ed 30:691
2. Ostwald W (1901) Phys Chem 37:385
3. Kabalnov AS, Shchukin ED (1992) Adv Colloid Interface Sci 38:69
4. Vollmer J, Vollmer D, Strey R (1997) J Chem Phys 107:3627
5. Puig JE, Perez-Luna VH, Perez-Gonzalez M, Macias ER, Rodrigueez BE, Kaler EW, (1993) Colloid Polymer Sci 271:114
6. Lezer NJ, Terrisse I, Bruneau F, Tokgoz S, Ferriera L, Seiller D, Grossiord JL (1997) J Control Release 45:1

7. Tadros ThF (1984) in Structure/Performance Relationships in Surfactants, Rosen MJ (ed), ACS Symp Ser, No. 253, American Chemical Society, Washington, D.C., p.154
8. Durbin DP, El-Aasser MS, Sudol ED, Vanderhoff JW (1979) J Appl Polym Sci 24:703
9. Chamberlain BJ, Napper DH, Gilbert RG (1982) J Chem Soc Faraday Trans 1 78:591
10. Barnette DT, Schork FJ (1987) Chem Eng Prog 83:625
11. Mouron D, Reimers JL, Schork FJ (1996) J Polym Sci Polym Chem Ed 34:1073
12. Wang S, Poehlein GW, Schork FJ (1997) J Polym Sci Polym Chem Ed 35:595
13. Chern CS, Chen TJ (1997) Colloid Polym Sci 275:546; ibid 275:1067
14. Chern CS, Chen TJ (1998) Colloids Surfaces A 138:65
15. Chern CS, Liou YC, Chen TJ (1998) Macromol Chem Phys 199:1315
16. Chern CS, Liou YC (1998) Macromol Chem Phys 199:2051
17. Chern CS, Liou YC (1999) Polymer 40:3763
18. Reimers JL, Skelland AHP, Schork FJ (1995) Polym React Eng 3:235
19. Reimers JL, Schork FJ (1996) J Appl Polym Sci 59:1833; ibid 60:251
20. Alduncin JA, Forcada J, Asua JM (1994) Macromolecules 27:2256
21. Alduncin JA, Asua JM (1994) Polymer 35:3758
22. Reimers JL, Schork FJ (1997) Ind Eng Chem Res 36:1085
23. Chern CS, Chen TJ, Liou YC (1998) Polymer 39:3767
24. Gardon JL (1968) J Polym Sci Part A-1 6:623; ibid 6:643
25. Harkins WD (1947) J Am Chem Soc 69:1428
26. Smith WV, Ewart RH (1948) J Am Chem Soc 70:3695
27. Smith WV, Ewart RH (1948) J Chem Phys 16:529
28. B. Jacobi (1952) Angew Chem 64:539
29. Priest WJ (1952) J Phys Chem 56:1077
30. Fitch RM, Tsai CH (1970) J Polym Sci Polym Lett Ed 8:703
31. Fitch RM, Tsai CH (1980) In: Fitch RM (ed), Polymer Colloids, Plenum, New York, p. 73
32. Hansen FK, Ugelstad J (1982) In: Piirma I (ed), Emulsion Polymerization, Academic Press, New York, p. 51
33. Billmayer FW (1991) Textbook of Polymer Science, 2nd ed Wiley, New York
34. Morgan JD, Kaler EW (1998) Macromolecules 31:3197
35. Gilbert RG (1995) Emulsion Polymerization. A Mechanistic Approach, Academic Press: London
36. Barton J, Capek I (1994) Radical Polymerization in Disperse Systems, E. Horwood, Chichester and Veda, Bratislava
37. Delgado J, El-Aasser MS Silebi CA, Vanderhoff J.W (1990) J Polym Sci Part A Polym Chem 28:777
38. Capek I, Barton J (1985) Makromol Chem 186:1297
39. Chang HC, Lin YY, Chern CS, Lin SY (1998) Langmuir 14:6632
40. Deryagnin BV, Lansau LD (1941) Acta Physicochim USSR 14:633
41. Verwey EJW, Overbeek JThG (1943) Theory of the Stability of Lyophobic Colloids, New York, Elsevier
42. Sato T, Ruch R (1980) Stabilization of Colloidal Dispersions by Polymer Adsorption, New York, Marcel Dekker
43. Napper DH (1983) Polymeric Stabilization of Colloidal Dispersions. London, Academic Press
44. Jansson L (1983) Master Thesis, Georgia Institute of Technology, Atlanta
45. Grimm WL, Min TI, El-Aasser MS, Vanderhoff JW (1983) J Colloid Interface Sci 94:531
46. El-Aasser MS, Misra SC, Vanderhoff JW, Manson JA (1977) J Coatings Tech 49:71
47. El-Aasser MS, Poehlein GW, Vanderhoff JW, (1977) Coatings and Plastics Preprints 37:2
48. Morton MJ (1954) J Colloid Interface Sci 9:300
49. Ugelstad J, Berge A, Ellingsen T, Schmid R, Nilsen TN, Mork PC, Stenstad P, Hornes E, Olsvik O (1992) Prog Polym Sci 17:87
50. Lifshitz IM, Slesov VV (1961) J Phys Chem Solids 19:35
51. Wagner C (1961) Ber Bunsenges Phys Chem 65:581

52. Kabalnov AS, Makarov KN, Pertzov AV, Shchukin ED (1989) J Colloid Interface Sci 138:98
53. Ugelstad J, Mork PC, Hansen FK, Kaggerud KH, Ellingsen T (1981) Pure and Appl Chem 53:323
54. Saethre B (1994) Mechanism and Kinetics in Minisuspension Polymerization of Vinyl Chloride, PhD. Dissertation, NTH, Trondheim
55. Lack CD, El-Aasser MS, Vanderhoff JW, Fowker FM (1985) Macro- and Microemulsions, Theory and Applications, Shah DO (ed), ACS Symp Ser No. 272, American Chemical Society, Washington, DC
56. Choi YT, El-Aasser MS, Sudol ED, Vanderhoff JW (1985) J Polym Sci Polym Chem Ed 23:2973
57. Ugelstad J, El-Aasser M, Vanderhoff JW (1973) J Polym Sci Polym Lett Ed 11:503
58. El-Aasser MS, Lack CD, Vanderhoff JW, Fowkes FW (1988) Colloids Surfaces 29:103
59. Tang PL, Sudol ED, Silabi CA, El-Aasser MS (1991) J Appl Polym Sci 43:1059
60. Higuchi WI, Misra J (1962) J Pharm Sci 51:459
61. El-Aasser MS, Lack CD, Choi YT, Min TI, Vanderhoff JW, Fowkes FM, (1984) Colloids Surfaces 12:79
62. Pithayanukul P, Pipel N (1982) J Colloid Interface Sci 89:494
63. Fowkes FM (1963) J Phys Chem 67:1982
64. Choi YT (1986) Formation and Stabilization of Miniemulsions and Latexes, Ph.D. Dissertation, Lehigh University, Bethlehem
65. MacRitchie F (1967) Nature (London) 170:1159
66. Vold RD, Mittal KL (1972) J Colloid Interface Sci 38:451
67. Lack CD, El-Aasser MS, Silebi CA, Vanderhoff JW, Fowkes FM (1987) Langmuir 3:1155
68. Goetz RJ, El-Aasser MS (1990) Langmuir 6:132
69. Goetz RJ, Khan A, El-Aasser MS (1990) J Colloid Interface Sci 137:795
70. Azad ARM, Ugelstad J, Fitch RM, Hansen FK (1976) Emulsion Polymerization, Piirma I, Gardon J (eds), ACS Symp Ser No. 24, American Chemical Society, Washington DC, p. 1
71. Zhou JS, Kamioner M, Dupeyrat M (1987) Microemulsion Systems, Rosano I, Henri L (eds), Marcel Dekker, p. 335
72. Delgado J (1986) Miniemulsion Copolymerization of Vinyl Acetate and n-Butyl Acrylate, Ph.D. Dissertation, Lehigh University, Bethlehem
73. Delgado J, El-Aasser MS, Vanderhoff JW (1980) J Polym Sci Polym Chem 24:861
74. Rodripuez VS (1988), Interparticle Monomer Transport in Miniemulsion Copolymerization, Ph.D. Dissertation, Lehigh University, Bethlehem,
75. Elsynski HA (1987) Batch and Semicontinuous Copolymerization of Vinyl Acetate and Methyl Acry]ate, M.S. Research Report, Lehigh University, Bethlehem
76. Miller CM, Sudol ED, Silebi CA, El-Aasser MS (1995) Macromolecules 28:2754
77. Ugelstad J (1980) Adv Colloid Interface Sci 13:101
78. Vijayendran BR (1979) J Appl Polym Sci 23:733
79. Wang S, Shork FJ (1994) J Appl Polym Sci 54:2164
80. Wang S, Schork FJ, Poehlein GW, Gooch JW (1996) J Appl Polym Sci 60:2069
81. Blythe PJ, Klein A, Sudol ED, El-Aasser MS (1999) Macromolecules 32:6952
82. Miller CM, Blythe PJ, Sudol ED, Silebi CA, El-Aasser MS (1994) J Polym Sci Part A Polym Chem 32:2365
83. Pelssers EGM, Cohen Stuart MA, Fleer GJ (1990) J Chem Soc Faraday Trans 86:1355
84. De Witt JA, van de Ven TGM (1992) Adv Colloid Interface Sci 42:41
85. Chern CS, Liou YC, Tsai WY (1996) J Macromol Sci Pure Appl Chem A33:1063
86. Capek I (1989) Makromol Chem 190:789
87. Capek I, Barton J (1985) Makromol Chem 186:1297
88. Capek I, Potisk P (1992) Polymer J 24:1037
89. Matsumoto A, Kodama K, Aota H, Capek I (1999) Eur Polymer J 35:1509
90. Miller CM, Sudol ED, Silebi CA, El-Aasser MS (1995) J Polym Sci Part A Polym Chem Ed 33:1391

91. Napper DH, Gilbert RG (1987) Makromol Chem Macromol Symp 10/11:503
92. Blythe PJ, Morrison BR, Mathauer KA, Sudol ED, El-Aasser MS (1999) Macromolecules 32:6944
93. Miller CM, Sudol ED, Silebi CA, El-Aasser MS (1995) Macromolecules 28:2765
94. Person LT, Louis PEJ, Gilbert RG, Napper DH (1991) J Polym Sci Polym Chem Ed 29:515
95. Russell G, Napper DH, Gilbert RG (1988) Macromolecules 21:2141
96. Ito K (1973) J Polym Sci Polym Chem Ed 13:401
97. Capek I, Riza M, Akashi M (1992) Polym J 24:959
98. Delgado J, El-Aasser MS, Silebi CA, Vanderhoff JW, Guillot J (1988) J Polym Sci Polym Phys Ed 26:1495
99. Delgado J, El-Aasser MS, Silebi CA, Vanderhoff JW (1989) J Polym Sci Polym Chem Ed 27:193
100. Nomura M, Harada M, Eguchi W, Nagata S (1976) in: Emulsion Polymerization, Piirma I, Gardon JL (eds), ACS Symp Seri, No. 24, American Chemical Society, Washington DC, p. 102
101. Friis N, Nyhagen L, (1973) J Appl Polym Sci 17:2311
102. Landfester K, Bechthold N, Tiarks F, Antonietti M (1999) Macromolecules 32:2679
103. Landfester K, Bechthold N, Forster S, Antonietti M (1999) Macromol Rapid Commun 20:81
104. Edelhauser H, Braitenbach JW (1959) J Polym Sci 35:423
105. Wang CC, Yu NS, Chen CY, Kuo JF (1996) Polymer 37:2509
106. Ugelstad J, Mork PC, Kaggerud KH, Ellingsen T, Berge A (1980) Adv Colloids Interface Sci 13:101
107. Shah DO (1971) J Colloid Interface Sci 37:744
108. Fontenot K, Schork FJ (1993) J Appl Polym Sci 49:633
109. Miller CM, Sudol ED, Silebi CA, El-Aasser MS (1995) Macromolecules 28:2772
110. Wu SQ, Schork FJ, Gooch JW (1999) J Polym Sci Polym Chem 37:4159
111. Reimers J, Schork FJ (1996) Polym React Eng 4:135
112. Capek I, Barton J, Tuan LQ, Svoboda V, Novotny V (1987) Makromol Chem 188:1723
113. Chern CS, Lin SY, Chang SC, Lin JY, Lin YF (1998) Polymer 39:2281
114. Chern CS, Lin SY, Chen LJ, Wu SC (1997) Polymer 38:1977
115. Chen LJ, Lin SY, Chern CS, Wu SC (1997) Colloids Surfaces A 122:161
116. Chern CS, Liou YC (1999) J Polym Sci Part A Polym Chem 37:2537
117. Capek I (1999) Adv Colloid Interface Sci 80:85
118. Matsumoto A, Murakami N, Aota H, Ikeda J, Capek I (1999) Polymer 40:5687
119. Matsumoto A, Kodama K, Aota H, Capek I (1999) Eur Polym J 35:1509
120. Alducin JA, Forcada J, Barandiaran, Asua J. M (1991) J Polym Sci Polym Chem Ed 29:1265
121. Capek I (2000) Polym J 32:670
122. Capek I (1999) Eur Polym Sci 36:255
123. Chern CS, Lin SY, Hsu TJ (1999) Polymer J 31:516
124. Capek I, Lin SY, Hsu TJ, Chern CS (2000) J Polym Sci Polym Chem Ed, 38:1477
125. Capek I, Chudej J (1999) Polym Bul 43:417
126. Chudej J, Capek I (2000) Polymer, in press
127. Garti N, Aserin A (1996) Adv Colloid Interface Sci 65:37
128. Coen E, Lyons RA, Gilbert RG (1996) Macromolecules 29:5128
129. Kusters JMH, Napper DH, Gilbert RG, German AL (1992) Macromolecules 25:7043
130. Harada M, Nomura M, Eguchi W, Nagata S (1991) J Chem Eng Jpn 4:54
131. Lansalot M, Farcet C, Charleux B, Vairon JP, Pirri R (1999) Macromolecules 32:7354
132. Landfester K, Willert M, Antonietti M (1999) Macromolecules 33:2370

Editor: Prof. K. Dušek
Received: August 2000

Recent Progress in Synthesis and Evaluation of Polymer-Montmorillonite Nanocomposites

Mukul Biswas, Suprakas Sinha Ray

Department of Chemistry, Presidency College, Calcutta-700 073, India
e-mail: dkmandal@cal.usnl.net.in

Abstract. The review aims at highlighting the significant developments in the field of poly-mer-montmorillonite clay based nanocomposites with specific focus on synthetic method-ologies used, characterization, and evaluation of relevant bulk properties of these compos-ites.

Synthetic procedures include (a) monomer impregnation of clay followed by polymeri-zation, (b) intercalation of monomers/polymers in clay, and (c) clay exfoliation techniques.

Structural and morphological characteristics of selective composite systems as studied by X-ray diffraction, scanning electron microscopy, and transmission electron microscopy have been discussed. Results of thermogravimetric analysis of various nanocomposites en-dorsing the general enhancement of the thermogravimetric stabilities of the polymer-based composites relative to the base polymer and application of differential scanning ca-lorimetry for studying delamination and cooperative relaxation phenomenon in these composites have been reviewed. Conductivity characteristics of various composites and manifestation of conductivity anisotropy in several systems have also been discussed.

The prospects of application of the montmorillonite-polymer nanocomposites as high performance materials in several applications have been discussed.

Advances in Polymer Science, Vol. 155
© Springer-Verlag Berlin Heidelberg 2001

List of Abbreviations

ANI	Aniline
ATBN	Amine Terminated Butadiene Acrylonitrile
ATBN-MMT	Amine Terminated Butadiene Acrylonitrile-Montmorillonite
BDMA	Benzyldimethylamine
BTFA	Borontrifluride monoethyl amine
C12A-SWy	$C_{12}H_{25}NH_3^+$ exchanged montmorillonite (Wyoming)
C18A-SWy	$C_{18}H_{37}NH_3^+$ exchanged montmorillonite (Wyoming)
CMS	Catalyst Modified Silicate
DGEBA	Diglycidylether of bisphenol A
DMSO	Dimethylsulfoxide
MDA	Methylene dianiline
MEEP	Poly[bis(methoxy-ethoxy) ethoxy phosphazene
MMA	Methylmethacrylate
MMT	Montmorillonite
MTS	Micatype silicate
NMA	Nadic maleic anhydride
NVC	N-Vinylcarbazole
OMTS	Organically modified micatype silicate
PANI	Polyaniline
PAN-KAO	Polyacrylonitrile-Kaolinite Composite
PEO	Polyethylene oxide
PLS	Polymer layered silicate
PMMA	Polymethylmethacrylate
PNVC	Poly(N-vinylcarbazole)
PPY	Polypyrrole
PS	Polystyrene
PTP	Polythiophene
PU	Polyurethane
PVA	Poly(vinyl alcohol)
PVP	Poly(N-vinylpyrrolidone)
PY	Pyrrole
S	Styrene
TP	Thiophene
VM-MMT	Vinylmonomer-Montmorillonite

1
Introduction

Clays are the most abundant inexpensive natural materials with high mechanical strength and high chemical resistance. They possess a layered structure, which can be exfoliated to yield an appreciable surface area which can be used for the adsorption of molecules [1, 2]. Clay polymer materials have received considerable research attention, since interactions between them influence the properties of both the clay and the polymer [3, 4].

This review is intended to highlight the major developments in this area during the last decade. The more important synthetic methodologies used for the preparation of various composites and their physicochemical characterization and evaluation of some distinctive bulk properties of these composites will be described in this article

In general, clay-polymer composites are of three categories [5–8]:

1. *Conventional composites*: in conventional composites, the clay tactoids exist in their original aggregated states with no intercalation of the polymer matrix into the clay lamellae.
2. *Intercalated composites*: in an intercalated composite the insertion of polymer into the clay structure occurs in a crystallographically regular fashion, regardless of the clay to polymer ratio. An intercalated nanocomposite is normally interlayered by only a few molecular layers of polymer and properties of the composite typically resemble those of ceramic materials.
3. *Exfoliated composites*: in an exfoliated nanocomposite, the individual clay layers are separated in a continuous polymer matrix by average distances that depend on loading. Usually, the clay content of an exfoliated composite is much lower than that of an intercalated nanocomposite.

2
Types of Polymers So Far Used for Composite Preparation with Clay (MMT)

The large variety of polymer systems that have so far been used in composite preparation can be conveniently classified as below.

2.1
Vinyl Polymers

These include the vinyl addition polymers derived from common monomers like methyl methacrylate [9], acrylonitrile [10–12], styrene [13], butadiene [14], 4-vinylpyridine [15], acrylamide [15], and tetrafluoro ethylene [16]. In addition, selective polymers like poly(vinyl alcohol) [17], poly(*N*-vinyl pyrrolidone) [18], poly(ethylene glycol) (PEG) [19], and others [20] have also been used.

2.2
Condensation (Step) Polymers

Several technologically important polycondensates have also been used in the composite preparations with MMT. These include nylon-6 [6, 21–25], poly(ϵ-caprolactone) [26], poly(ethylene oxide) [27], poly(dimethyl siloxane) [8], epoxy resins [28, 29], and polyurethanes [30, 31]

2.3
Specialty Polymers

In addition to the above-mentioned conventional polymers, several interesting developments have taken place in the preparation of nanocomposites of MMT with some specialty polymers including the N-heterocyclic polymers like poly (N-vinylcarbazole) (PNVC) [32, 33], polypyrrole (PPY) [34, 35], and polyaromatics such as polyaniline (PANI) [36–38]. PNVC is well known for its high thermal stability [39] and characteristic optoelectronic properties [40–43]. PPY and PANI are known to display electric conductivity [44–46]. Naturally, composites based on these polymers should be expected to lead to novel materials [47, 48].

Since PPY and PANI are intractable polymers, a good deal of research attention has been directed to prepare processable dispersions of these polymers to facilitate their application. However, for such preparation nanodimensional inorganic oxides including SiO_2 [49–51], MnO_2 [52], ZrO_2 [53, 54], and SnO_2 [55] have been used suitably as particulate dispersants. Some work has also been initiated to prepare MMT-based nanocomposite dispersions of these polymers, including PNVC [32, 35, 38].

3
Synthetic Methodologies

The general procedures so far used for obtaining clay-polymer nanocomposites may be classified as below.

3.1
Impregnation with Monomers Followed by Polymerization

This involves impregnation of the clay with vinyl monomers followed by their polymerization induced by a free radical. Conventional monomers which have been used include acrylonitrile [10–12], methyl methacrylate [9], styrene [13], and tetrafluoroethylene [16]. Blumstein [9] prepared monomolecular layers of MMA on clay by saturating the clay with monomer followed by removing the excess monomer by repeated washing with nonpolar hydrocarbons. Complexes of clay with 25% MMA produced inserted polymers, while those with more than 25% MMA produced both insertion as well as surface-adsorbed polymers.

Table 1. Retention of PNVC in the MMT layer

Entry no.	Weight of MMT (g)	% Conversion to PNVC[a,b]	% PNVC Extracted with benzene[b]	% PNVC Unextracted with benzene[b]	% PNVC Retained per g of MMT
1	0.02	55.00	53.70	1.30	2.80
2	0.03	56.00	54.04	1.96	3.30
3	0.05	70.00	66.80	3.20	4.20
4	0.10	73.30	63.30	10.00	7.30
5	0.20	74.00	54.00	20.00	7.50
6	0.30	47.30	41.05	6.25	1.00
7	0.10	94.00	92.50	1.50	–[c]
8	0.10	94.00	92.60	1.40	–[c]

[a] In each case (entry nos. 1–6) polymerization was carried out at 50 °C for 1 h; concentration of NVC solution=0.1 mol/l; volume of NVC/benzene solution=5 ml
[b] Data represent average of three sets of experiments
[c] Conducted with preformed PNVC (0.09 gm PNVC in 5 ml benzene, conversion 94%)

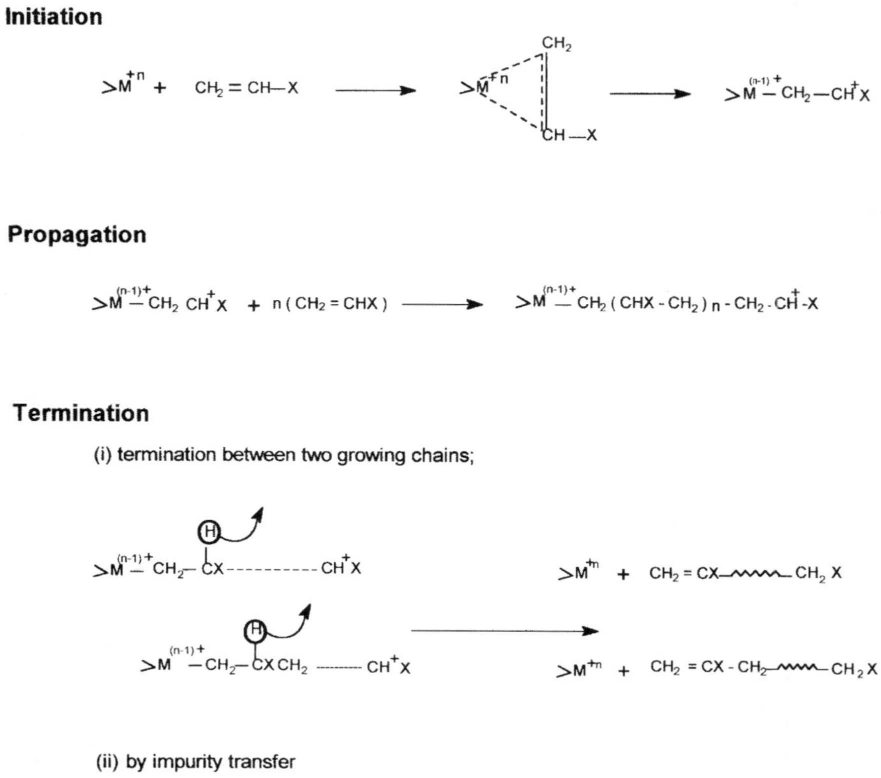

Initiation

Propagation

Termination

(i) termination between two growing chains;

(ii) by impurity transfer

Scheme 1. Initiation of NVC by transition metal oxides

Biswas and Sinha Ray [32] recently prepared a PNVC-MMT composite by direct polymerization of NVC (solid) in presence of MMT without the use of any free radical initiator. Melt polymerization of NVC in MMT (at 70 °C) as well as solution polymerization of NVC (benzene) in the presence of MMT at 50 °C resulted in the formation of PNVC. Repeated benzene extraction of PNVC-MMT mass resulted in the formation of a composite from which a residual PNVC could not be removed, while all the surface-adsorbed PNVC was extracted with benzene (Table 1). XRD analysis confirmed intercalation of PNVC in the MMT lamellae.

As with NVC-Zeolite (13X and SK-500) systems, the initiation in the NVC-MMT system was suggested [56–58] to be cationic involving Brφnsted acid sites in MMT arising from the dissociation of interlayer water molecules coordinated to the exchangeable cations [59]. Yet another possibility, especially with NVC, was that the transition metal oxides Fe_2O_3/TiO_2 present in MMT could also lead to cationic initiation of NVC (Scheme 1) [32, 59]:

Biswas and Sinha Ray [35] subsequently reported that direct interaction of MMT with PY led only to ca. 5% yield of PPY in 3 h while ANI could not be polymerized by MMT. Such a trend is possibly not surprising since NVC is relatively more susceptible to cationic polymerization compared to the latter monomers.

3.1.1
Effect of FeCl₃ on Polymerization/Nanocomposite Formation in N-Vinylcarbazole-MMT System

$FeCl_3$ is an efficient Lewis acid initiator for the cationic polymerization of monomers like NVC [60]. It is also an efficient dopant for PNVC enhancing its bulk conductivity [61]. Accordingly, in the study of the NVC-MMT polymerization/nanocomposite formation system, the addition of $FeCl_3$ was considered to be interesting. Results of a recent study [33] (Table 2) indicated that in NVC-

Table 2. Loading of PNVC in PNVC-MMT composite

Entry No.	Weight of MMT (g)	Weight of FeCl₃ (gm)	% Conversion to PNVC[a,b]	% PNVC present per g of composite
1	0.100	–	73.60	10.00
2	0.100	0.007	86.00	12.90
3	0.102	0.015	94.00	19.50
4	0.100	0.029	95.60	18.60
5	0.100	0.050	~100	[c]

[a] In each case, polymerization was conducted at 50 °C for 1 h with 0.1 mol/l NVC in benzene and total volume of the reaction system was 5 ml
[b] Data represent average of three sets of experiments
[c] Polymer gelled out in the medium as a separate layer and no loading of PNVC occurred in the MMT layers

Table 3. Composite formation in PY-MMT system

Entry No.	Weight (g) in initial feed			% PPY per g of MMT in composite	Conductivity (S/cm)
	MMT	FeCl$_3$	PY		
1	0.101	–	0.097	20.30	1.3×10^{-5}
2	0.102	–	1.74	22.35	–
3	0.202	–	1.74	10.70	–
4	0.101	0.03	1.74	52.33	–
5	0.201	0.025	1.74	52.66	–
6	0.202	0.05	1.74	81.96	2.5×10^{-5}
7	0.202	0.10	1.74	81.74	5.9×10^{-5}
8	0.102	0.035	0.0	30.73	3.9×10^{-5}
9	0.200	0.10	1.74	22.00	–
10	0.200	0.42	1.74	113.00	8.3×10^{-5}
11	0.200	0.68	1.74	118.40	9.0×10^{-5}
12	0.200	0.80	1.74	285.00	26.0×10^{-5}

Table 4. Some typical data for ANI-MMT [$(NH_4)_2S_2O_8$] polymerization/nanocomposite formation system

Entry No.	Weight of MMT (g)	Weight of Oxidant (g)	Volume of ANI (ml)	Time of polymerization (h)	% yield of PANI	% PANI per g of composite	Conductivity (S/cm)
1	0.1	0.2	0.2	3	36.17	73.10	–
2	0.1	0.4	0.2	3	78.10	60.50	0.3×10^{-3}
3	0.1	0.4	0.2	6	79.30	61.00	–
4	0.1	0.4 +0.2[a]	0.2	3+3	100.00	66.66	–
5	–	0.4	0.2	3	78.00	–	3×10^{-1}
6	0.2	0.4	0.2	3	81.10	44.62	0.7×10^{-3}
7	0.3	0.4	0.2	3	76.00	33.41	1×10^{-3}
8	0.4	0.4	0.2	3	72.10	25.26	5×10^{-3}
9	0.4	1.0	0.2	3	100.00	32.85	1.1×10^{-3}
10	0.4	1.0	0.4	3	100.00	48.93	5×10^{-3}
11	0.4	1.0	0.6	3	93.20	58.50	9×10^{-3}

[a] Late addition of catalyst

MMT polymerization/nanocomposite formation system addition of FeCl$_3$ increased the percentage loading of PNVC in the composite.

Polymerization vis-a-vis nanocomposite formation in PY-MMT-water and ANI-MMT-water systems was possible after using FeCl$_3$ and $(NH_4)_2S_2O_8$ [49] as oxidant respectively in the two systems. Results of recent studies in these systems are presented in Tables 3 and 4, respectively.

The preparation of PNVC-MMT nanocomposites deserves some specific comments. Hitherto, reported preparations of stable aqueous dispersions of intractable polymer systems was accomplished with water soluble monomers like

PY or ANI which would be polymerized in aqueous particulate dispersions of nanodimensional oxides, etc., in presence of $FeCl_3$ or $(NH_4)_2S_2O_8$ and the polymer precipitating out in the aqueous medium would eventually encapsulate the particulate dispersants and produce processable dispersions. However, NVC monomer is insoluble in water and so the common procedures could not be applied here. MMT-based composites of PNVC were produced during precipitation of benzene soluble polymer along with the nanosized MMT particles by methanol addition and in this process the precipitating PNVC particles encapsulated the MMT particles leading to nanocomposites of PNVC encapsulated MMT particles [33]. The same procedure was successfully applied for the preparation of $MnO_2/ZrO_2/SiO_2$ based PNVC nanocomposites which were readily dispersible in the aqueous media [51–53].

3.2
Intercalation of Monomers/Polymers into the Clay Lamellae

The general idea underlying the preparation of layered clay-polymer intercalates follows the simple rules of ion exchange. In most cases, the synthesis involves either intercalation of a polymer from solution or a suitable monomer followed by subsequent polymerization. But for more technologically important polymers, both approaches are limited since neither a suitable monomer nor a compatible polymer host solvent system is always available.

3.2.1
Solution Radical Polymerization Technique

In this technique [59] the grafting of polymers onto the clay interlayers depends on the swelling of the modified organophilic clay promoted by the solvent. Swelling is primarily due to solvation of the interlayer organic cations and the inclusion of a vinyl monomer between the layers of clay which can be maximized by the use of an appropriate solvent. The swelling phenomena of clays result from a balance between the interlayer cohesive forces and the attractive forces between the solvent and the intercalated cations. Hence the interlayer distance can be increased markedly in the solvents which have strong attractive forces to the intercalated monomer. Thus, the role of the solvent is to improve wetting and penetration of the monomers in the clay interlayers. Hence, the solvent used must have attractive forces strong enough to increase the interlayer distance and must also be a good solvent for the monomer molecules between the clay layers prior to the onset of the polymerization.

For example, vinyl monomer-montmorillonite intercalate (VM-MMT), able to swell and disperse in organic solvents, was prepared by exchanging the mineral cations of MMT by vinylbenzyltrimethylammonium chloride. The resulting VM-MMT material rendered the mineral organophilic and having polymerizable moieties directly bonded to the lamellar surface of the mineral. Radical polymerization of S between the interlayers of 5, 10, 25, and 50 wt % of VM-MMT

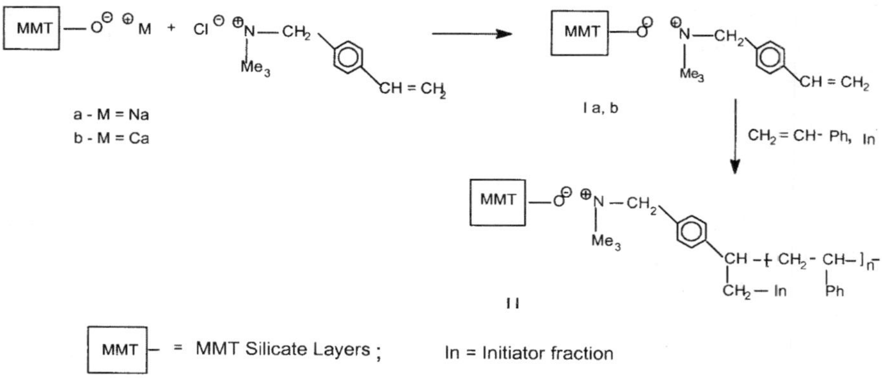

Scheme 2. Preparation of VM-MMT intercalates and their free radical polymerization with styrene

in selected solvents resulted in PS-MMT intercalate materials (Scheme 2) [62, 63].

3.2.2
Direct Intercalation of Polymers from Solution

This technique has been used for the intercalation of vinyl polymers in MMT. Akelah et al. [64] reported the preparation of butadiene acrylonitrile (ATBN)-MMT nanocomposites by using simple rules of ion exchange in solution. Thus the amine end groups of the ATBN polymer were converted into the corresponding onium salt, which was then allowed to be exchanged with the interlayer cation of MMT. Accordingly, a weighed amount of ATBN was dispersed in a mixed solvent composed of 1:1 (V/V) mixture of DMSO and water with a very small amount of concentrated HCl (8 N) (Scheme 3) [63].

The ATBN dispersion was added dropwise to a suspension of MMT (~4 g) in 200 ml water and stirred for 24 h at room temperature using a propeller stirrer. The product was filtered, washed with water several times, and collected by a filter press, followed by drying at 80 °C in a convection oven, then in a vacuum oven at 100 °C for 24 h in each case giving 94% yield.

Organophilic PMMA-MMT and PS-MMT supported catalysts have been prepared by grafting the copolymers of PMMA [9] or PS [62, 63] with CMS that contained 2% ammonium salt groups onto MMT layers through direct cation exchange process. The remaining chloromethyl groups of the grafted copolymers have been modified to produce different onium salts (Scheme 4). The structural composition and catalytic activity of the onium moieties supported on PS-MMT and PMMA-MMT materials have been investigated in different nucleophilic substitution reactions under the effects of several factors governing the reaction rates, such as the composition of the polymer backbone, nature of active moiety, degree of loading, nature of the solvent, and reaction temperature.

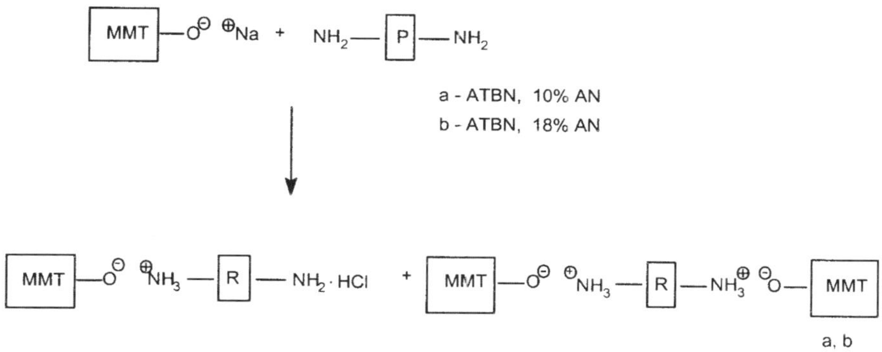

$$MMT-O^{\ominus}\ ^{\oplus}Na\ +\ NH_2-P-NH_2$$

a - ATBN, 10% AN
b - ATBN, 18% AN

$$MMT-O^{\ominus}\ ^{\oplus}NH_3-R-NH_2\cdot HCl\ +\ MMT-O^{\ominus}\ ^{\oplus}NH_3-R-NH_3^{\oplus}\ ^{\ominus}O-MMT$$

a, b

Scheme 3. Preparation of rubber-MMT materials

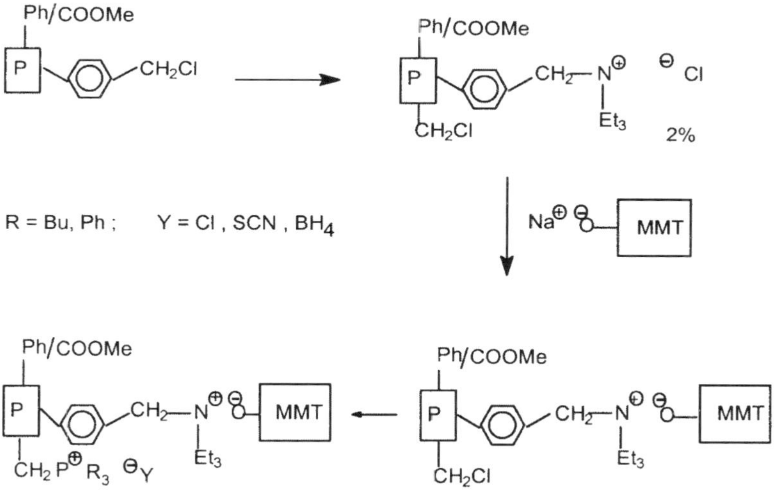

Ph/COOMe

$$P-\bigcirc-CH_2Cl \longrightarrow$$

Ph/COOMe

$$P-\bigcirc-CH_2-N^{\oplus}\ ^{\ominus}Cl$$
$$\ \ \ \ |\ \ \ \ \ \ \ \ \ \ \ \ Et_3$$
$$CH_2Cl$$

2%

R = Bu, Ph ;   Y = Cl , SCN , BH$_4$

$$Na^{\oplus}\ ^{\ominus}O-MMT$$

Ph/COOMe

$$P-\bigcirc-CH_2-N^{\oplus}\ ^{\ominus}O-MMT$$
$$\ \ \ \ |\ \ \ \ \ \ \ \ \ \ \ \ \ \ Et_3$$
$$CH_2P^{\oplus}R_3\ ^{\ominus}Y$$

$$\longleftarrow$$

Ph/COOMe

$$P-\bigcirc-CH_2-N^{\oplus}\ ^{\ominus}O-MMT$$
$$\ \ \ \ |\ \ \ \ \ \ \ \ \ \ \ \ \ \ Et_3$$
$$CH_2Cl$$

Scheme 4. Preparation of PS-MMT and PMMA-MMT supported catalyst

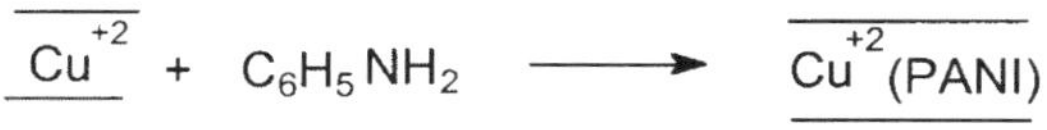

$$\overline{Cu^{+2}}\ +\ C_6H_5NH_2\ \longrightarrow\ \overline{Cu^{+2}\ (PANI)}$$

Scheme 5. Intercalative polymerization of ANI in Cu-fluorohectorite

Recently, Mehrotra and Giannelis [36, 37] reported the intercalative polymerization of ANI in Cu-fluorohectorite oriented films from the liquid phase (or vapor phase) resulting in a rapid color change from light blue to purple. The PANI-silicate hybrid, after thorough washings with acetone, contained 17.5 wt % polymer. Gallery Cu^{+2} ions (Scheme 5), introduced by an ion exchange process, served as the oxidation centers for the polymerization of ANI. The reaction could be represented by the scheme, where the horizontal lines identified the layer structure.

During polymer intercalation from solution, a relatively large number of solvent molecules need to be desorbed from the host to accommodate the incoming polymer chain. The desorbed solvent molecules gain on translational degree of freedom, and the resulting entropic gain compensates for the decrease in conformational entropy of the confined polymer. Thus the driving force for the polymer intercalation from solution is the entropy gained by desorption of the solvent molecules due to the gain in translational freedom of many desorbed molecules from the clay galleries. The unfavorable loss of conformational entropy associated with the intercalation of the polymer can be overcome by maximization of the number of polymer-clay intercalations. The intercalated chains must establish numerous segment-clay contacts to offset the large decrease in conformational energy that opposes intercalation. These interactions impede the rotational and translational motion of the polymer. The specificity of the polymer can be enhanced by eliminating the competing clay solvent and polymer-solvent interaction.

3.2.3
Direct Polymer-Melt Intercalation

Recently, Vaia et al. [8] reported a new process for direct polymer intercalation based on a predominantly enthalpic mechanism. By maximization of the number of polymer host interactions, the unfavorable loss of conformational entropy associated with intercalation of the polymer can be overcome leading to new intercalated nanostructures. They also reported that this type of intercalated polymer chain adopted a collapsed, two-dimensional conformation and did not reveal the characteristic bulk glass transition. This behavior was qualitatively different from that exhibited by the bulk polymer and was attributed to the confinement of the polymer chains between the host's layers. These types of materials have important implications not only in the synthesis and property areas, where ultrathin polymer films confined between adsorbed surfaces are involved. These include polymer filler interactions in polymer composites, polymer adhesives, lubricants, and interfacial agents between immiscible phases.

The organosilicate host was a derivative of montmorillonite, a mica-type layered silicate (MTS) which possessed the same structural characteristics as the well known minerals talc and mica [65] comprising two tetrahedral silica sheets fused to an edge shared octahedral sheet of either aluminum or magnesium hydroxide. Stacking of the layers by weak dipolar or Van der Waals forces would lead to interlayers or galleries between the layers. The galleries were normally occupied by cations which balanced the charge deficiency generated by isomorphous substitution.

Organically modified silicates were produced by a cation exchange reaction between the silicate and an alkylammonium salt [66–68]. The presence of alkylammonium cation in the galleries rendered the hydrophilic MTS organophilic.

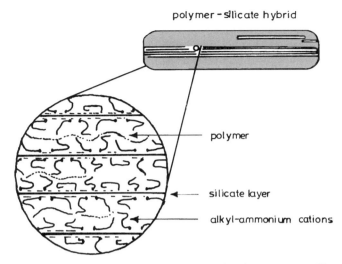

polymer-silicate hybrid

polymer

silicate layer

alkyl-ammonium cations

Fig. 1. Schematic illustration of polymer chains intercalated in an organosilicate

The direct intercalation involved the mixing of an organo-silicate with a polymer, pressing the mixture into a pellet, and heating at the appropriate temperature which was always greater than the bulk glass transition temperature of polymer ensuring the presence of a polymer melt. For example, PS [8] was intercalated by mixing 0.25 g of PS with 0.75 g of alkylammonium MMT and heating the pellet in vacuum at 165 °C. In addition to PS, other polymers of varying degrees of polarity and crystallinity, including poly(vinylidene fluoride), poly(ethylene oxide) (PEO), poly(ε-caprolactone), and poly(dimethyl siloxane) could be directly intercalated in organically modified silicate [4, 69]. A schematic illustration of polymer chains intercalated in an organosilicate is shown in Fig. 1 after Vaia et al. [8].

A literature survey also revealed that the above-mentioned composites were inaccessible by solution intercalation techniques.

Attempts to intercalate PS in the organosilicate from solution (toluene) resulted in intercalation of the solvent instead of the polymer. Therefore, the direct intercalation enhanced the specificity for the polymer intercalation by eliminating the competing host-solvent and polymer solvent interaction. Vaia et al. [8] also reported that PS could be deintercalated by suspending the hybrid in toluene. The d-spacing of the deintercalated silicate sample was identical to that of the starting organosilicate. Gel permeation chromatography showed that the displaced PS was identical to the starting polymer, suggesting that the intercalated polymer retained its molecular weight and polydispersity and that the process was completely reversible.

Condesation polymer-MMT based nanocomposites were formed, while polymerizing bifunctional monomers in the presence of organophilic modified clay containing active groups capable of forming a bond between the clay and the formed condensation polymer. For example, the condensation polymerization of ε-caprolactam in the presence of clay-aminolauric or aminocaproic acid intercalates was reported to give a nylon 6-clay hybrid [8, 19, 70, 71].

The advantages [8] of direct intercalation are considerable. Foremost, direct intercalcation is highly specific for the polymer leading to new hybrids, which were previously inaccessible. In addition, the absence of a solvent makes direct intercalation an environmentally sound and economically advantageous method. Finally, intercalate hybrids represent ideal systems to study polymers in a restricted, two-dimensional geometry by conventional techniques.

3.3
Clay Exfoliated Polymer Composites

MMT exfoliated by polymeric molecules exhibit a wide range of novel physical properties. One specially interesting system recently reported by Toyota researchers [72] was based on the exfoliation of $[H_3N(CH_2)_{11}COOH]^+$-MMT in a semicrystalline nylon-6 matrix. In this context, Fujiwara and Sakamota [73] chose a specific organo clay-polymer system to demonstrate the swelling of the clay layers in a thermoplastic polyamide matrix. $[H_3N(CH_2)_5COOH]^+$-MMT was dispersed in ε-caprolactone and the mixture was heated at 250 °C to form a clay/nylon-6 composite. Very recently, Wang and Pinnavaia [28] reported the preparation of a new type of clay-polymer composite by spontaneous self-polymerization of an epoxy resin, the *diglycidylether of bisphenol A (DGEBA)*, with the structure I.

$$\overset{O}{\underset{CH_2-CHCH_2}{\triangle}} O-\left[\!\!\left\langle\bigcirc\right\rangle\!-C(CH_3)_2-\left\langle\bigcirc\right\rangle-OCH_2\overset{OH}{\underset{CHCH_2}{|}}O\right]_{\!\!n}\!\!-\left\langle\bigcirc\right\rangle-C(CH_3)_2-\left\langle\bigcirc\right\rangle-OCH_2-\overset{O}{\underset{CH-CH_2}{\triangle}}$$

The commercially available resin I, EPON-828 (Shell) n=0 (88%), n=1 (10%), and n=2 (2%) [74, 75] and with an average molecular weight of 378, was used for that reaction. The reaction of EPON-828 resin with H^+-, NH_4^+-, $[H_3N(CH_2)_{n-1}CH_3]^+$- (II), $[H_3N(CH_2)_{n-1}CO_2H]^+$- (III), $[H_3N(CH_2)_nNH_2]^+$- (IV), or $[H_3N(CH_2)_nNH_3]^{+2}$- (V) exchanged forms of MMT (where n=6 and 12) at temperatures in the range 200–300 °C resulted in the polymerization of the epoxide and the concomitant exfoliation of acid form of MMT at elevated temperatures. Their preparation procedure involved the addition of a desired amount of ion exchanged clay to a weighed amount of epoxy resin EPON-828 and the clay-epoxide mixture was magnetically stirred in a 250-ml beaker at 75 °C for 30 min. The beaker was sealed with aluminum foil, and the temperature was then raised at a rate of ~20 °C/min to the clay exfoliation epoxide polymerization temperature. The liquid to powder transformation of the mixture and the concomitant formation of the clay-epoxide composite took place within 1 min.

Primary amines and carboxylic acids are commonly used as curing agents for epoxy resins owing to the reaction of amino and carboxylic groups with the epoxide ring according to Eqs. (1)–(3):

$$H_3\overset{+}{N}(CH_2)_{n-1}\overset{O}{\overset{\|}{C}}\text{-OH} \quad + \quad CH_2\text{—}CHR \quad \longrightarrow \quad H_3\overset{+}{N}(CH_2)_{n-1}\overset{O}{\overset{\|}{C}}\text{—OCH}_2\overset{OH}{\overset{|}{CHR}}$$

(1)

$$R^1\overset{+}{NH_3} \quad + \quad CH_2\text{—}CHR \quad \longrightarrow \quad R^1 H_2\overset{+}{N}CH_2\overset{OH}{\overset{|}{CHR}}$$

(2)

$$R^1 H_2\overset{+}{N}CH_2\overset{OH}{\overset{|}{CHR}} \quad + \quad CH_2\text{—}CHR \quad \longrightarrow \quad R^1 \overset{+}{HN}(CH_2\overset{OH}{\overset{|}{CHR}})_2$$

(3)

However, at the onium ion clay loading (typically 5 wt %) the fraction of epoxide functional groups capable of reacting with the amino and carboxylic moieties on the clay-exchange cations was limited by stoichiometry to the range 1.4–2.8%. Thus the protonated aminocarboxylic acid and primary diamine cations on the exchange sites of MMT acted primarily as acid catalysts rather than as curing agents. The acid catalytic role of the onium ions was verified by the fact that the same type of polymerization-exfoliation reaction was observed for the H^+, NH_4^+- and $[H_3N(CH_2)_{n-1}CH_3]^+$- exchange forms of the clay. Consequently, the observed primary polymerization was the formation of a polyether according to Eqs. (4) and (5), where $R'OH$ was an epoxide monomer possessing one or two hydroxyl groups:

$$RH\overset{+}{\overset{HO}{C}}\text{—}CH_2 \quad + \quad RCH\text{—}CH_2 \quad \longrightarrow \quad \overset{RHC}{\underset{H_2C}{\diagdown}}\overset{+}{O}\text{—}CHRCH_2\,OH \quad (4)$$

$$\overset{RHC}{\underset{H_2C}{\diagdown}}\overset{+}{O}\text{—}CHRCH_2\,OH + n\,RCH\text{—}CH_2 \quad \longrightarrow \quad \overset{RHC}{\underset{H_2C}{\diagdown}}\overset{+}{O}\text{—}(CHRCH_2O)_nCHRCH_2\,OH$$

(5)

In the above process reported by Wang and Pinnavaia [28], the product of the high temperature curing reaction was an intractable powder rather than a continuous solid epoxy matrix. In this background, Messersmith and Giannelis [29] reported the preparation of an MTS-epoxy nanocomposite which fulfilled the above requirement and was processed using conventional epoxy curing agents at temperatures significantly lower than those previously reported by Lan and Pinnavaia [6]. The resulting composite exhibited molecular dispersion of the silicate layers in the epoxy matrix, good optical clarity, and significantly improved dynamic mechanical properties compared to the neat epoxy resin.

An organically modified mica-type silicate (OMTS) was prepared by an ion-exchange reaction from MMT and bis(2 hydroxy ethyl) methyl tallow – alkyl ammonium chloride as shown in Eq. (6), where R' was predominantly an octadecyl chain with smaller amount of lower homologues and R'' a methyl group:

$$Na^+_- MTS + (HOCH_2CH_2)_2 R'R''N^+Cl^- \longrightarrow (HOCH_2CH_2)_2 R'R''N^+_- MTS + NaCl$$

(6)

The dry OMTS powder was added with stirring to DGEBA and cured by addition of either nadic maleic anhydride (NMA), boron trifluoride monoethyl amine (BTFA), benzyldimethyl amine (BDMA), or methylene dianiline (MDA). The OMTS-DGEBA mixture was held at 90 °C with stirring for 1 h and then sonicated for 1–2 min. Following sonication, samples were cooled, curing agent was added with thorough mixing, and then loaded into disposable syringes. Samples were centrifuged in the syringes for 30 s at 3000 rpm to remove bubbles, and then dispersed into rectangular Teflon molds with dimensions 20 mm by 10 mm by 1.5 mm thick or cast as free standing films with thickness of 0.1–0.3 mm. All samples were cured at 100 °C for 4 h, 150 °C for 16 h, and 200 °C for 12 h.

Nanolayer reinforcement of elastomeric polyurethane (PU) was achieved by Wang and Pinnavaia [28] by a novel reaction. To disperse MMT nanolayers in the PU matrix, the hydrophilic inorganic exchange cations of the native mineral (Wyoming) MMT with unit cell formula of $Na_{0.80}[Mg_{0.86}Al_{3.14}]$-$(Si_{8.00})O_{20}(OH)_4]$ (SWy-2) were replaced with more organophilic alkyl ammonium ions $C_{12}H_{25}NH_3^+$ (C12A) and $C_{18}H_{37}NH_3^+$ (C18A). MMT, exchanged thus, was easily solvated by ethylene glycol, polyethylene glycol, polypropylene glycol, and voranol glycerol propoxylates. Representative PU nanocomposites were formed by adding a Rubinate methylene diphenyl diisocyanate prepolymer (MW 1050) to a mixture of organoclay and a polyol – glycerol propoxylate. The organoclay was preintercalated with the polyol by stirring with a magnetic stirrer at 50 °C for 12 h in a sealed container. The prepolymer was added to the clay-polyol intercalate and stirred at 70 °C for a definite time before degassing at 95 °C in a vacuum oven. The mixture was subsequently poured into a stainless steel mold for curing at 95 °C for 10 h in an inert atmosphere.

Hutchison et al. [76] prepared poly[bis(methoxy-ethoxy) ethoxyphosphazene] (MEEP)-NaMMT nanocomposite in a composition of 1 MEEP repeat unit $(PN(CH_3OC_2H_4OC_2H_4O)_2)$ to 2 clay Si_8O_{20} units by adding 1/2 equivalent of MEEP [76] dissolved in CH_3CN to a slurry of NaMMT.

Recently, Okari and Lerner [34] reported a general method for the incorporation of intractable, but preformed polymers into MMT. Accordingly, they reported the inclusion of PPY and PTP into MMT by a polymer latex-exfoliated host colloid interaction. Colloidal PPY and PTP were prepared using the method described by DeArmit and Armes [77]. An oxidant and a stabilizer, sodium dodecyl benzenesulfonate, were dissolved in deionized water at ambient temperature to form a turbid solution and known volumes of PY or TP were syringed into the reaction mixture and the solution stirred for 7 h. The resulting suspension was centrifuged, yielding a black solid, consisting of nanoparticles, that could be readily dispersed into deionized water under mechanical stirring. The latex was not dried fully, as this was observed to result in the irreversible agglomeration of the nanoparticles. The product was stripped of the stabilizer, residual oxidant, and other soluble components through repeated cycles of centrifugation and redispersion in deionized water. The final wash was tested and found to be free of residual monomer.

The polymer-clay nanocomposites were synthesized by exfoliating a known amount of clay in water and then mixing this with a very small amount of polymer. Nanocomposite particles flocculated from the solution within 1–2 h. The mixture was stirred for 12–24 h and the nanocomposites were isolated by centrifugation followed by washing and drying for at least 24 h in vacuo.

4
Characterization of Montmorillonite-Based Nanocomposites

Useful information about the structural and morphological characteristics of the various MMT-polymer composites has been available through X-ray diffraction (XRD), scanning electron microscopic (SEM), and transmission electron microscopic (TEM) analyses. Some relevant results of several interesting clay-based polymer composites are highlighted below.

4.1
X-ray Diffraction Characterization

4.1.1
PNVC-MMT Nanocomposite System

Figure 2 indicates the manifestation of a peak at 14.6 Å in the d_{001} reflection of PNVC-MMT nanocomposites which confirms the intercalation of PNVC in the MMT lamellae. In addition another peak at 9.8 Å was also observed which is commonly ascribed to the d_{001} spacing in unintercalated MMT.

Interestingly, for PNVC-MMT nanocomposite produced in the NVC-MMT ($FeCl_3$ impregnated) system, the peak at 14.6 Å was missing, implying no intercalation of PNVC in the MMT although a part of the PNVC was found to be retained in the MMT layer [33]. Similar observations were also made in PPY-MMT and PANI-MMT nanocomposite systems [35,38] where no evidence for the expansion of the d_{001} spacing of MMT could be obtained.

4.1.2
ATBN-MMT Composite System

Akelah et al. [64] reported the expansion of the d_{001} spacing in MMT in the amine terminated butadiene/acrylonitrile-montmorillonite system (ATBN-MMT) to 14 Å (Fig. 3) and of the span between the lamella surface to 5 Å, implying horizontal packing of the polymer molecule as shown in Fig. 3. In their patent work, which covered every possible class of polymers, a Toyota group of re-

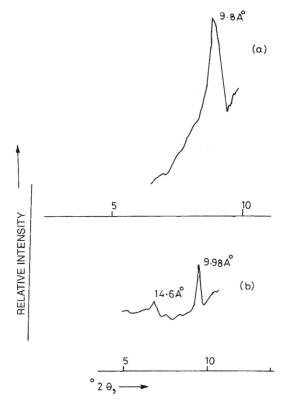

Fig. 2a,b. XRD patterns of: **a** MMT powder; **b** PNVC-MMT composite powder

searchers [72] disclosed that the XRD analysis of ATBN-MMT also exhibited the disappearance of the d_{001} spacing and the uniform dispersion of MMT in the liquid butadiene. In another report [63] for ATBN intercalate treated with an excess of polymer, the silicate layers were found to be dispersed in the molecular level. However, recent [63] WAXD measurements further verified that interlamellar spacing remained in the range of 13.7 Å after intercalative polymerization of ε-caprolactone. A model proposed for representing the morphological hierarchy consistent with the XRD and TEM analyses findings is reproduced in Fig. 4 after Akelah et al. [64]

4.1.3
PAN-KAO Composite System

Sugahara et al. [11] confirmed from the appearance of a broad XRD peak at 13–14 Å the formation of the PAN-KAO intercalation complex during the polymerization of AN between the kaolinite layers.

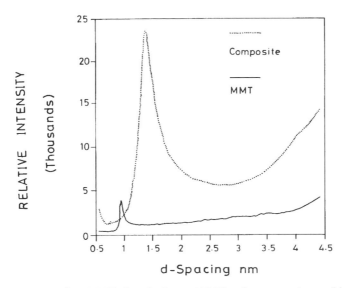

Fig. 3. XRD patterns of Na.MMT after drying at 100 °C and compression molded samples of ATBN-MMT

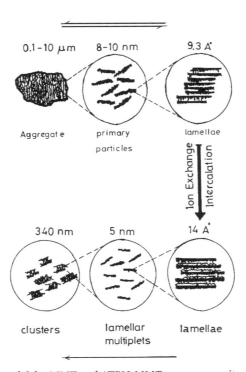

Fig. 4. Hierarchical model for MMT and ATBN-MMT nanocomposite

4.1.4
PS-MMT Composite System

Recently, Giannelis et al. [8] studied the XRD pattern of PS hybrid without heating and after 2, 5, 15, and 25 h of heating (Fig. 5); asterisks indicate the positions of basal reflections from the pristine organosilicate. The XRD pattern before heating contained peaks characteristic of the pristine organosilicate (d_{001}= 2.25 nm and d_{002}=1.27 nm). After heating, the intensity of the diffraction peaks corresponding to the pristine organosilicate was progressively reduced while a set of new peaks corresponding to the PS-organosilicate intercalated hybrid appeared. After 25 h, an XRD pattern corresponding predominantly to the intercalated hybrid could be observed. Likewise, the intercalation of higher molecular weight PS400 occurred in a similar manner, albeit at a slower rate. In both cases, the resulting gallery height increase of 0.7 nm corresponded to a monolayer of nearly collapsed chain (Fig. 1). Incidentally, the XRD pattern of the organosilicate, heated under identical conditions in the absence of the polymer, remained unchanged.

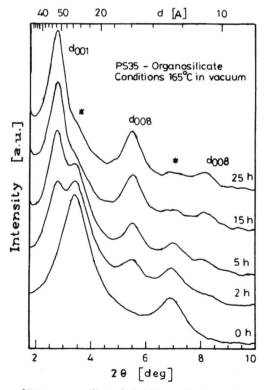

Fig. 5. XRD patterns of P35-organosilicate hybrid heated to 165 °C

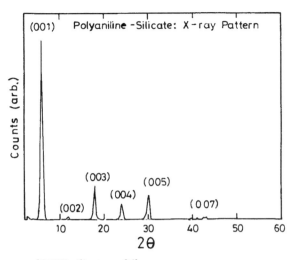

Fig. 6. XRD patterns of PANI-silicate multilayers

4.1.5
PANI-MTLS Composite System

The XRD patterns of PANI-silicate multilayers reported by Mehrotra and Gian-nelis [36] are shown in Fig. 6. Seven (001) harmonics were observed correspond-ing to a primary repeat unit (d- spacing) of 14.9 Å. The difference of 5.3 Å from the corresponding 9.6 Å for the silicate framework was in agreement with inter-calation of single chains of PANI in the galleries of the host.

4.1.6
Delamination Studies

Wang and Pinnavaia [28] studied by XRD analysis the delamination of the MMT clay in the polymerized epoxy resin (structure I). As shown by the powder pat-terns in Fig. 7, $[NH_3(CH_2)_{11}COOH]^+$-MMT remained crystalline over the tem-perature range 25–229 °C. However, no clay diffraction peaks were observed for a 5:95 (w/w) clay polyether nanocomposite formed from $[NH_3(CH_2)_{11}COOH]^+$-MMT at 229 °C (Fig. 7c). Only very diffuse scattering, characteristic of the amor-phous polyether, appeared in the XRD pattern of the composite. The absence of the 17.0 Å peak for $[NH_3(CH_2)_{11}COOH]^+$-MMT suggested that the clay particles were exfoliated and the 9.6 Å-thick clay layers were dispersed at the molecular level. XRD analyses were also used to study the rate of progress of OMTS disper-sion during mixing with epoxy resin, DGEBA, and subsequent curing reactions [29]. Figure 8 shows XRD patterns of dry OMTS and the uncured OMTS-DGEBA mixture. The XRD pattern of the OMTS powder showed a primary silicate (001) reflection at $2\theta=4.8°$, with a low intensity shoulder at roughly $2\theta=5.8°$. The main

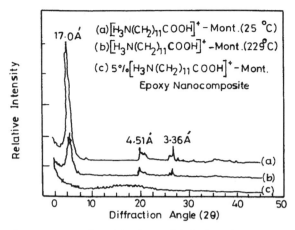

Fig. 7. XRD powder patterns for (a) freeze-dried $[NH_3(CH_2)_{11}COOH]^+$-MMT, (b) $[NH_3(CH_2)_{11}COOH]^+$-MMT freeze-dried and then heated at 229 °C, and (c) clay-polyether nanocomposite containing 5 wt % $[NH_3(CH_2)_{11}COOH]^+$-MMT

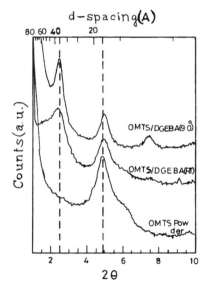

Fig. 8. XRD patterns of OMTS powder and uncured OMTS-DGEBA mixture. Top scan was obtained at room temperature following heating of OMTS-DGEBA mixture at 90 °C for 1h

silicate reflection in OMTS corresponded to a d-spacing of 17 Å which implied an increase of approximately 7 Å from the Van der Waals gap of NaMMT [78]. Following mixing of OMTS and DGEBA at room temperature, an additional reflection appeared at $2\theta=2.5°$ which corresponded to the intercalated OMTS. As

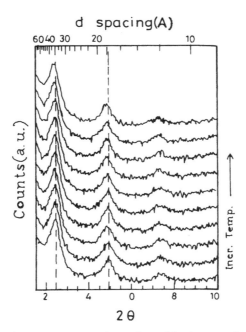

Fig. 9. XRD patterns of OMTS-DGEBA mixture heated in situ to various temperatures

is widely known, OMTS can readily intercalate various small organic molecules from either the vapor or liquid phase [3, 10, 30, 36, 37, 79–84]. The second peak at $2\theta=5°$ corresponded to the coexistence of unintercalated ($d_{001}=17$ Å) and intercalated ($d_{002}=17.5$ Å) OMTS. The persistence of some intercalated OMTS could also be seen by the small remnant shoulder at $2\theta=5.8$ Å. In contrast, mixing of OMTS and DGEBA at 90 °C resulted in only DGEBA intercalated OMTS ($d_{001}=35$ Å) with no residual OMTS peak observed, as shown in the top trace of Fig. 8. The reflections observed at $2\theta=2.5°$, $4.9°$, and $7.6°$ corresponded to the (001), (002), and (003) reflections of the DGEBA intercalated phase, respectively. In situ XRD experiments were used to determine the exact structure of the resin mixtures at elevated temperatures.

Figure 9 represents a series of XRD scans of the OMTS-DGEBA mixture previously heated at 90 °C taken at various intervals between room temperature and 150 °C. The low temperature scans exhibited three orders of reflections indicating the existence of DGEBA intercalated OMTS with $d_{001}=36$ Å. With increasing temperature, a gradual increase in d_{001} from 36 Å to approximately 39 Å was observed, although the constant intensity of the peaks suggested that little or no delamination occurred at or below 150 °C. It was observed that the choice of curing agent was critical in determining delamination and optical clarity. A survey of the common epoxy curing agents revealed an unexpected fact that many curing agents resulted in little or no increase in layer separation, resulting in com-

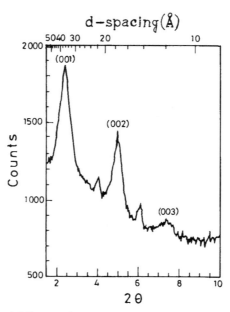

Fig. 10. XRD pattern of fully cured OMTS-DGEBA-MDA composite containing 2 vol.% OMTS

posites with a silicate d-spacing of 30–40 Å or less. An example of this behavior is shown in Fig. 10 for MDA cured OMTS-DGEBA composite. Interestingly, all bifunctional primary and secondary amine curing agents used were found to have this effect and resulted in opaque composites, in contrast to the transparent composites following delamination of OMTS.

Unlike the primary and secondary amines, a number of curing agents (NMA, BDMA, BTFA, and combinations thereof), were found to result in OMTS delamination during heating of the reaction mixture. Figure 11 represents the in situ XRD scans of the OMTS-DGEBA-BDMA mixture illustrating the delamination of OMTS on heating from room temperature to 150 °C. The sample was prepared by mixing OMTS and DGEBA in a vial at 90 °C, cooling to room temperature, and mixing in BDMA immediately before transferring to the diffractometer chamber. The mixing of BDMA into the OMTS-DGEBA resin at room temperature resulted in an intercalated system with $d_{001} = 39$ Å. But in the absence of a curing agent, heating of the OMTS-DGEBA-BDMA mixture resulted in substantial attenuation of the peak at $2\theta = 2.3°$. This peak almost disappeared by 150 °C (top of Fig. 11), with only a trace remaining at $2\theta = 3°$. The virtual disappearance of the OMTS (001) reflections clearly endorsed delamination of OMTS.

XRD analysis of completely cured nanocomposite samples also lacked the silicate (001) reflections as shown in Figs. 12 and 13 for OMTS-DGEBA-BMDA and OMTS-DGEBA-NMA, respectively. The absence of the silicate (001) reflections in the cured nanocomposites showed that the delamination and dispersion of

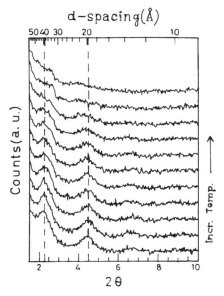

Fig. 11. XRD patterns of OMTS-DGEBA-BDMA mixture (4 vol.%) heated in situ to various temperatures.

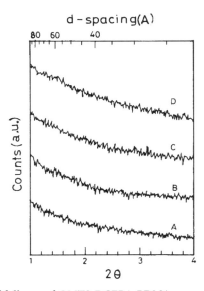

Fig. 12. XRD patterns of fully cured OMTS-DGEBA-BDMA nanocomposites

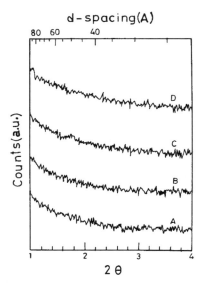

Fig. 13. XRD patterns of fully cured OMTS-DGEBA-NMA nanocomposites

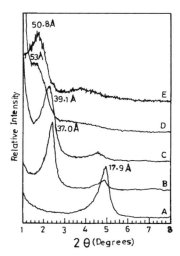

Fig. 14. XRD patterns of C18A-Swy MMT: (A) unsolvated clay; (B) solvated by polyol. The remaining curves are for solvated clay at different times: (C) 95 °C for 10 min; (D) 95 °C for 25 min; (E) 95 °C for 10 h

the silicate layers within the epoxy matrix were retained after complete curing of the epoxy.

Wang and Pinnavaia [31] were able to prove for the organo-MMT-PU nano-composite systems that the expansion of the organo-clay increased with increasing curing time (Fig. 14) at 95 °C. Gallery expansion beyond the initial value for polyol solvation was evident from the appearance of a broad diffraction peak at low angle (D in Fig. 14). Significantly, the intergallery polymerization contributed to the dispersal of the nanolayers. The basal spacing (E in Fig. 14) at equilibrium was found to be less than the spacing expected for bilayer orientation of onium ions which indicated the formation of a layered silicate intercalate in the fully cured nanocomposite. The presence of Braggs reflection was suggestive of the stacking of the nanolayers in crystallographical order in the polymer matrix.

4.2
Thermogravimetric Analysis

Thermogravimetric analyses have confirmed the general enhancement in the stability of the various polymer-clay nanocomposites relative to the base polymers and also provided useful information on the extent of the polymer loading in such composites. The thermogravimetric characteristics of several typical systems are highlighted below.

4.2.1
MMT-based Nanocomposites of PNVC, PPY, and PANI

Biswas and Sinha Ray [32] obtained weight-loss vs temperature data for MMT (Fig. 15b) revealing that unintercalated MMT registered a total weight-loss of 13.24% due to probable loss of volatile impurities in MMT. Figure 15a, indicates that the PNVC-MMT intercalate lost 19.59% of mass which comprised loss of intercalated PNVC along with the volatile impurities present in the system. These data implied that the intercalated MMT had a PNVC loading of 6.35% per unit mass of MMT which compared reasonably well with the value of 7.3% obtained gravimetrically (Table 1).

The weight-loss vs temperature data for PANI-MMT systems (Table 5) revealed that while PANI [38] decomposition was complete at 571 °C onwards, the composite underwent 64% weight loss at 950 °C. Subtracting from this figure the 13% lost due to loss of MMT at 999 °C. a PANI loading of 51% was realized, which coincided convincingly with the gravimetrically observed PANI loading of 49% in the composite (entry 10, Table 4).

The TG data in Table 5 further confirm that the polymer-MMT nanocomposites were thermally more stable than the respective base polymers.

Table 5. Weight loss vs temperature data for MMT based nanocomposites of PNVC and PANI

Materials	% Weight loss [temperature (°C)]		
MMT	8.3 (190.9)	12.2 (682)	13.2 (999.7)
PNVC	8 (400)	75 (500)	complete (700)
PANI	15.42 (134.90)	24.13 (259)	complete (571.90)
PNVC-MMT composite	10.(454)	17.7 (604)	19.5 (999)
PANI-MMT composite	12.88 (117.80)	42.98 (487.60)	64.09 (950)

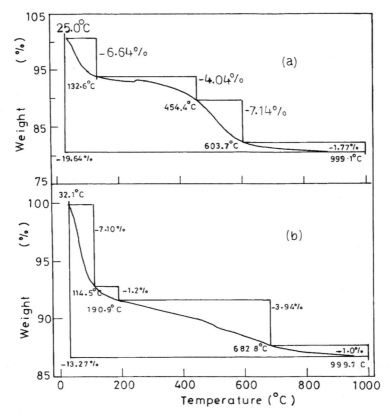

Fig. 15 a,b. Weight loss vs temperature thermograms: **a** PNVC-MMT nanocomposite; **b** MMT powder.

4.3
Differential Scanning Calorimetric Studies

4.3.1
PS-Organosilicate Composite System

DSC studies have been successfully used to obtain evidence of intercalation. Giannelis et al. [8] reported (Fig. 16) that the thermal behavior of intercalated PS-organosilicate was different from the thermal characteristics of the bulk polymer. Both the pure polymer and the physical mixture of PS-organosilicate clearly exhibited the characteristic glass transition at 96 °C but the intercalated hybrid did not show any in the range 50–150 °C.

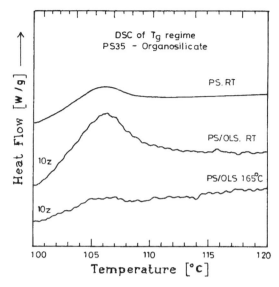

Fig. 16. DSC traces of pure PS, a physical mixture of PS-organosilicate and PS-intercalated organosilicate heated at 20 °C/min. under nitrogen

Table 6. PDT values and thermodynamic data for clay-polyether nanocomposites formed from bifunctional onium ion MMT[a]

Interlayer cation	Initial basal spacing (Å)	PDT[b] (°C)	Heat of reaction (J/g)	Heat of polymerization[c] (kJ/m)
$[H_3N(CH_2)_{11}COOH]^+$	17.0±0.1	229±1	572±16	228±6
$[H_3N(CH_2)_5COOH]^+$	13.3±0.0	248±1	565±06	225±2
$[H_3N(CH_2)_{12}NH_3]^{2+}$	13.4±0.1	271±1	566±08	225±3
$[H_3N(CH_2)_6NH_3]^{2+}$	13.1±0.1	273±2	568±07	226±3
$[H_3N(CH_2)_{12}NH_2]^+$	13.5±0.0	281±2	563±07	224±3
$[H_3N(CH_2)_6NH_2]^+$	13.2±0.1	287±2	557±03	222±2

[a] The clay:polymer composition was 5:95 (w/w)
[b] PDT is the onset temperature for epoxide polymerization-clay delamination at a heating rate of 20 °C/min
[c] Heat of reaction for two epoxide equivalents

4.3.2
Organoclay-Epoxide Nanocomposites

Wang and Pinnavaia [28] obtained the onset temperature for the polymerization-delamination (PDT) and overall heats of reaction for polymerization of EPON-828 in the galleries of MMT containing the bifunctional onium ions listed in Table 6. In a typical system of 5:95 (w/w) mixture of $[H_3N(CH_2)_{11}COOH^+]$-

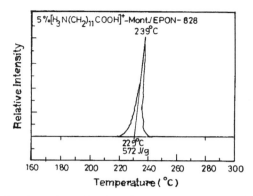

Fig. 17. DSC curve for the polymerization of EPON-828 epoxy resin in the presence of 5 wt % $[NH_3(CH_2)_{11}COOH]^+$-MMT at a heating rate of 20 °C

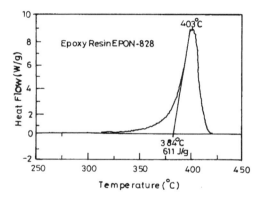

Fig. 18. DSC thermogram (20 °C/min) for the uncatalyzed self-polymerization of epoxy resign EPON-828

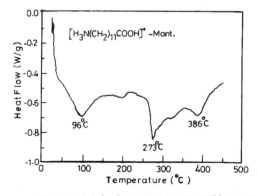

Fig. 19. DSC thermogram (20 °C/min) for $[NH_3(CH_2)_{11}COOH]^+$-MMT

MMT and EPON-828, the DSC curve (Fig. 17) indicates that the spontaneous clay exfoliation epoxide polymerization occurred at an onset temperature of 229 °C. On the basis of the integrated peak area, the heat of reaction for the composite was 572 J/g. DSC curves for the neat epoxy resin and the pristine $[H_3N(CH_2)_{11}$-COOH$]^+$-MMT are shown for comparison in Figs. 18 and 19 respectively. In the absence of a catalyst, self polymerization of the neat resin occurred at a much higher onset temperature of 384 °C but the heat of reaction (611 J/g) was comparable to that observed for the corresponding nanocomposite (572 J/g), when corrected for the presence of clay (572/0.95) or 602 J/g.

4.3.3
OMTS-DGEBA Curing Process

A particularly interesting case [29] involved the use of the curing agent BDMA, which catalyzed the homopolymerization of DGEBA, and was also capable of catalyzing the reaction between hydroxyl groups of the alkylammonium ions and the oxirane rings of DGEBA [85, 86]. For example, increasing the temperature of the OMTS-DGEBA-BDMA and DGEBA-BDMA mixtures from 20 °C to 250 °C at slow rates (5 °C/min) resulted in little difference in the curing behavior between the composite and the unmodified epoxy. At higher rates, however, a difference in curing behavior manifested. Figure 20 shows the DSC scans of the OMTS-DGEBA-BDMA and DGEBA-BDMA curing reaction at a scanning rate of 10 °C/min, indicating a strong exotherm associated with curing between 100 °C and 150 °C for OMTS-DGEBA-BDMA. DSC scan of the DGEBA-BDMA mixture showed a considerably smaller exotherm over the same temperature range which suggested two possibilities: (a) OMTS played a catalytic role in the base catalyzed homopolymerization of DGEBA, or (b) the reaction proceeded by an

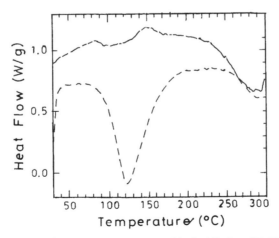

Fig. 20. DSC curing scans of OMTS-DGEBA-BDMA (*dashed line*) and DGEBA-BDMA (*solid line*)

altogether different mechanism in the presence of OMTS. Equation (7) represents the base-catalyzed oxirane ring-opening reaction between the hydroxyl groups of OMTS and DGEBA, resulting in the formation of **II** (an OMTS-glycidyl ether of bisphenol A oligomer). **II** could subsequently react with the free DGEBA

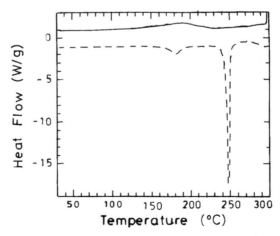

(7)

via similar base-catalyzed oxirane ring opening to build up the cross-linked epoxy network. It was interesting to note that the temperature at which curing occurred (approximately 100 °C as shown by the exotherm in Fig. 20) corresponded to the same temperature at which the delamination of OMTS occurred. The temperature coincidence of curing and delamination appeared reasonable, since delamination exposed the hydroxyl groups of the alkylammonium chains in the interlayer to DGEBA and BDMA.

The participation of the hydroxylated OMTS alkyl-ammonium ion in the curing reaction was more clearly illustrated with the OMTS-DGEBA-NMA system. Interestingly, full curing of the DGEBA-NMA mixture did not occur in the absence of OMTS, regardless of the heating rate. During dynamic curing of this formulation, two distinct exotherms were observed (Fig. 21) – a weak one at

Fig. 21. DSC curing scans of OMTS-DGEBA-NMA (*dashed line*) and DGEBA-NMA (*solid line*)

180 °C followed by a strong exotherm at 247 °C. A possible reaction was that of the OMTS hydroxyl groups with NMA to form the monoester, III (Eq. 8) [85, 87]. Nascent carboxylic groups of III could subsequently react with the epoxide resulting in the formation of the diester, IV, according to Eq. (9). Further reaction of III with DGEBA resulted in an epoxy net work formation:

(8)

(9)

4.3.4
Clay-PEO Nanocomposites

Krishnamoorti et al. [88] were able to demonstrate that local and global dynamic behavior of confined polymer chains were markedly different from the bulk. DSC experiment with intercalated PS or PEO/MMT chains revealed the absence of any thermal transition corresponding to glass or melting transition of these polymers.

Thermally stimulated current (TSC) technique was also deployed in studying cooperation relaxation phenomenon; TSC was applied in two modes – 'global' and 'thermal sampling' – with intercalated PEO. In the latter mode, TSC measurements were sensitive to cooperative relaxation from a minor fraction of the overall relaxing species [89–92]. Figure 22 shows the temperature dependence of apparent E_{act} (as determined with TSC) for 100% PEO, 20% PEO melt-intercalate, and 0% PEO in the MMT control pellet. Cooperation glass transition-like motions were assigned to regions of departure from $\Delta S^{\pm} = 0$ [89–92] prediction. For 20% intercalate a distinct peak in E_{act} was not observed and, instead, a broad transition starting at about the nominal PEO T_g ranging up to about 60 °C was observed [93–95].

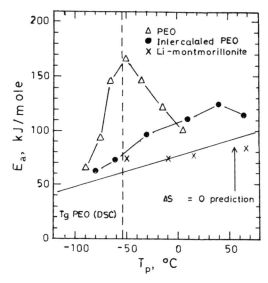

Fig. 22. TSC measurement curve

4.4
Transmission Electron Micrographic (TEM) Analysis

4.4.1
ATBN-MMT Composite

From TEM studies on ultramicrotone sections of ATBN-MMT composites Ake-lah et al. [64] observed that the mineral domains were ellipsoids with sizes rang-ing from 100 nm to 1000 nm with an average of 250 nm (see Fig. 4)

4.4.2
PNVC/PPY/PANI-MMT Nanocomposite Systems

Biswas and Sinha Ray [32, 33, 35, 38] reported that the TEM image of MMT (Fig. 23) exhibited elongated string-like particles fused together in short-chains. TEM images of PNVC-MMT, PPY-MMT, and PANI-MMT under different condi-tions are presented in Fig. 24a–f. Table 7 indicates the particle size distributions as evaluated from the TEM photographs.

Figure 24b indicates that MMT particles were not uniformly coated with the PPY particles apparently because of the poor yield of PPY in the system. The TEM photograph of a nanocomposite with high PPY loading (Fig. 24c) revealed relatively thicker and non-uniform coating of PPY on the MMT particles. In fact, a clear separation of well formed spherical particles was not distinguishable as

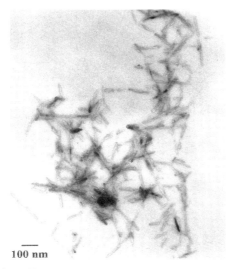

Fig. 23. TEM image of MMT

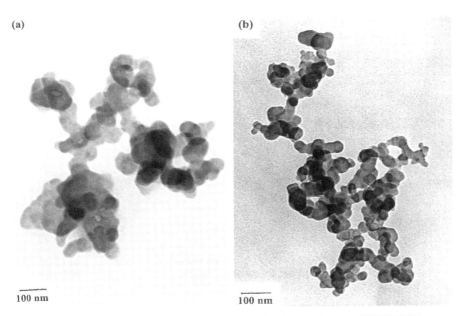

Fig. 24. **a** TEM image of PNVC-MMT nanocomposite. **b** TEM image of PPY-MMT nano-composite with low PPY loading

(c)   (d)

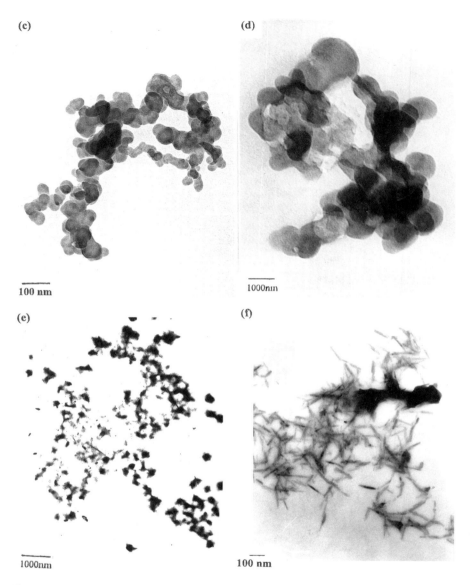

Fig. 24. **c** TEM image of PPY-MMT nanocomposite with high PPY loading. **d** TEM image of PANI-MMT nanocomposite. **e** TEM image of PANI-MMT nanocomposite at high dilution. **f** TEM image of PANI-MMT mechanical mixture (1:1, wt./wt.)

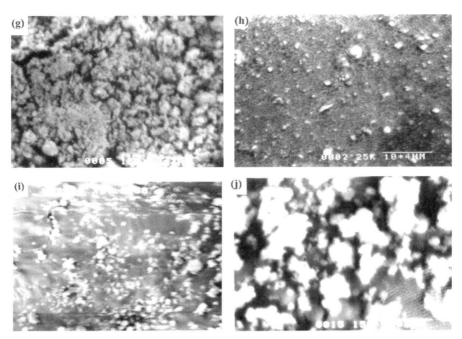

Fig. 24. g SEM image of PANI-MMT nanocomposite. **h** SEM image of extremely dilute colloids of PANI-MMT nanocomposite. **i** SEM image of PANI-MMT mechanical mixture (1:1, wt/wt). **j** SEM image of pure PANI powder

Table 7. Particle sizes of PNVC-MMT, PPY-MMT, and PANI-MMT composites

Composites	Particle size (nm)
PNVC-MMT	20–42
PPY-MMT	18–32
PANI-MMT	300–400

in the case shown in Fig. 24b with the low PPY loaded composite. A similar morphological feature was confirmed by Butterworth et al. [96] for the PPY-silica nanocomposite system.

The formation of the nanocomposite was clearly accompanied by a distinct change in the TEM pattern of MMT. For example, spherical PANI particles appeared to cluster around the stringy MMT particles [38]. The TEM photograph for a mechanical mixture of PANI and MMT (Fig. 24f) clearly revealed the non-uniform coverage of MMT particles by the PANI moieties as distinct from the composite (Fig. 24d,e).

Armes et al. contended in recent studies [49, 50, 55, 96, 97] that the PANI and PPY colloidal silica based nanocomposite particles were characterized by a typical raspberry morphology – where ultrafine silica particles were present not only on the surface but also distributed throughout the interior of the agglomerates. As the water-soluble monomers like ANI or PY started polymerizing in the aqueous medium, the insoluble polymers seemingly "glued" [49, 50, 96, 97] the silica particles to the microaggregates of the polymer-silica particles. In the PANI-MMT nanocomposite system a similar situation also appeared to prevail.

4.4.3
Polyether-MMT Nanocomposites

Unambiguous evidence for the delamination of MMT clay in the polymerized epoxy matrix was obtained from TEM analysis [26]. TEM images for a 5 wt % $[H_3N(CH_2)_{11}COOH]^+$-MMT polyether nanocomposite distinctly revealed that the micron-sized clay tactoids were expanded by the polymer into an accordion-like arrangement with interlayer spacing ranging up to 2000 Å. Evidently, the epoxide monomers appeared to polymerize into the gallery regions of MMT. Further, the clay-bound polyether particles became phase-segregated from the extra gallery regions during the polymerization process. For several epoxy-silicate-nanocomposites [6] from OMTS/epoxy resin (DGEBA) and curing agents (BDMA), TEM studies confirmed that the exfoliation of the silicate layers in the BDMA composites could be clearly distinguished by the manifestation of individual silicate layers of 1 nm thickness embedded in the epoxy matrix.

4.5
Scanning Electron Micrographic (SEM) Analysis

While TEM analyses of the various MMT-based nanocomposites yielded useful information about the morphological features, the SEM analyses were not particularly effective. For example, Akelah et al [64] failed to identify the mineral domains of the ATBN-MMT composite at 20,000 magnification although TEM analyses confirmed portraits of nanodimensional domains.

For the MMT based composites of PNVC [32] and PPY [35] the SEM analyses likewise were not particularly informative. Figure 24g–j compares the SEM images of (g) PANI-MMT nanocomposite, (h) PANI-MMT nanocomposite of an extremely dilute colloid, (i) 1:1 (w/w) mechanical mixture of PANI and MMT, and (j) pure PANI powder respectively. The SEM image of the dilute colloidal (100 ppm) composite clearly indicated the formation of nearly spherical particles covered with PANI particles. The SEM image of the undiluted colloid (g) revealed the highly dense packing of the composite particles. The SEM for the 1:1 (w/w) mechanical mixture of PANI and MMT revealed a morphology distinct from that of PANI or of the composite. In general, the SEM images of the composite exhibited a visual resemblance to the raspberry morphology exhibited by PANI-silica composite reported by Stejskal et al. [49].

5
Evaluation of Mechanical Properties

5.1
OMTS-DGEBA-BMDA Composite

Messersmith et al. [27] were able to show that pertinent mechanical properties of the clay/polymer matrix could be modified significantly. Shown in Fig. 25 are the temperature dependences of the tensile strength modulus ε, and tan δ of OMTS-DGEBA-BMDA composite containing 4% silicate by volume and the DGEBA-BMDA epoxy without any silicate. The shift and broadening of the tan δ peak to higher temperatures suggested an increase in the nanocomposite T_g and broadening of the glass transition. Broadening and an increase of T_g were reported [98–100] in other organic-inorganic nanocomposite systems and were generally attributed to restricted segmental motions near the organic-inorganic interface. Chemical bonding at the interface of the silicate-epoxy matrix could also lead to hindered relaxational mobility in polymer segments near the interface which could lead to broadening and increase of T_g. The nanocomposite exhibited a plateau modulus about 4.5 times higher than the unmodified epoxy (a change to be regarded as substantial in view of a silica loading of about 4% by volume).

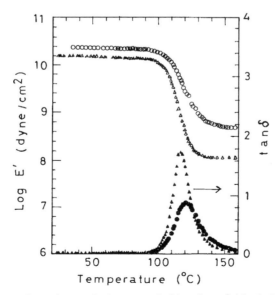

Fig. 25. Temperature dependence of ε (*open symbols*) and tan δ (*shaded symbols*) for fully cured DGEBA-BDMA (Δ,$\blacktriangle$) and OMTS-DGEBA-BDMA ($\bigcirc$,$\bullet$) containing 4 vol.% MTS

5.2
PLS Nanocomposites

The unprecedented mechanical properties of the polymer layered silicate (PLS) nanocomposites were first demonstrated by a group at the Toyota Research Center in Japan using nylon nanocomposites [72]. A doubling of the tensile modulus and strength could be achieved for the nylon layered silicate nanocomposite containing as little as 2 vol.% of inorganic material. Krishnamoorti et al. [88] demonstrated that the mass transport of PS chain into the layered silicates was unhindered by the confinement of the chains in the composite materials. The diffusion coefficients and their temperature dependence were similar to those in the bulk. Nanorheological features of several intercalated, exfoliated, and end-tethered exfoliated PLS nanocomposites revealed interesting differences. Figure 26 shows the steady shear viscosity as a function of shear rate for a series of intercalated hybrids of a copolymer of dimethyl siloxane and diphenylsilane with dimethylditallowmontmorillonite with varying levels of silica loadings [88]. The viscosity was enhanced considerably at low shear rates and increased monotonically with increasing silicate loading at a fixed shear rate. For a series of poly(dimethyl siloxane)-based delaminated hybrids (Fig. 27), steady shear viscosity showed an increase with respect to that for the pure polymer at low shear rates but still obeyed Newtonian behavior even at the highest silica loadings examined. With increasing shear rate, the polymer and the exfoliated hybrids exhibited shear thinning with the shear thinning character setting in at lower shear rates with increasing silica content.

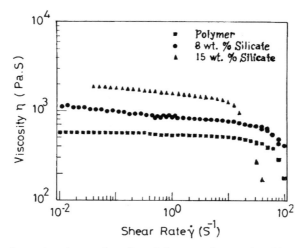

Fig. 26. Steady-shear viscosity as a function of shear rate for a series of intercalated hybrids of a copolymer of dimethylsiloxane and diphenylsiloxane with dimethylditallow MMT with varying levels of silicate loading at T=28 °C

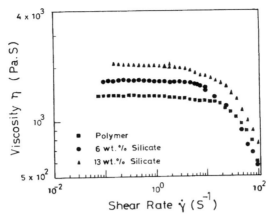

Fig. 27. Steady-shear viscosity as a function of shear rate for a series of delaminated hybrids of poly(dimethylsiloxane) with dimethylditallow-MMT with varying levels of silicate loading at T=28 °C

5.3
Epoxy-Clay Nanocomposites

Lan and Pinnavaia [6] showed that hybrid organic-inorganic composites exhibited mechanical properties superior to those for their separate composites. Dependence of tensile strength and modulus of epoxy nano-clay composites on the chain length of the clay-intercalated alkylammonium ions is shown in Fig. 28. The presence of the organoclay substantially increased both the tensile strength and modulus relative to the pristine polymer. The mechanical properties increased with any exfoliation in the order:

$$CH_3(CH_2)_7\text{-}NH_3^+\text{-}<CH_3(CH_2)_{11}NH_3^+\text{-}<CH_3(CH_2)_{17}NH_3^+\text{-}$$

Reinforcement of the epoxy-clay nanocomposites was also dependent on the clay loading as shown in Fig. 29. Thus for epoxy $CH_3(CH_2)_{17}NH_3^+$-MMT nanocomposite system, the tensile strength and modulus increased nearly linearly with clay loading. More than a tenfold increase in strength and modulus could be realized by the addition of ca. 15 wt % of the exfoliated organoclay.

Interestingly, for epoxide-*m*-phenylene diamine-clay nanocomposites the tensile strength and modulus were marginally improved relative to the base polymer. In contrast, for low T_g epoxide amine system under discussion-the reinforcement provided by the clay was much more significant due to shear deformation and stress-transfer to the platelet particles. Additionally, platelet alignment under strain might also contribute to the improved performance of clays exfoliated in a rubber matrix as compared to the glassy matrix. Figure 30 proposes a model after Lan and Pinnavaia [6] for the fracture of a glassy and a

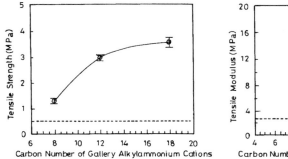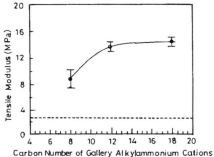

Fig. 28. Dependence of tensile and strength modules of epoxy-organoclay composites on the chain length of the clay-intercalated alkylammonium ions. The *dashed lines* indicate the tensile strength and modulus of the polymer in absence of clay

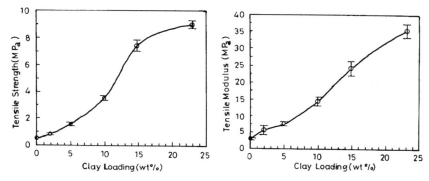

Fig. 29. Dependence of tensile strength and modulus on clay loading for epoxy-$CH_3(CH_2)_{17}NH_3^+$-MMT nanocomposite

rubbery matrix. In a glassy matrix, alignment of clay particles upon applying strain was minimal and blocking of fracture by the exfoliated clay was less efficient.

5.4
Organoclay-PU Nanocomposites

Typical stress-strain curves for the pristine PU elastomer and the organoclay-PU nanocomposites are shown in Fig. 31. Clearly, the nanolayers, even when aggregated in the form of intercalated tactoids, strengthened, stiffened, and toughened the matrix. At a loading of only 10 wt.% of the organo-clay, the strength, modulus, and strain at break were all increased by more than 100%. These tensile properties further increased with increasing load of the silicate monolayers.

A : Glassy Matrix

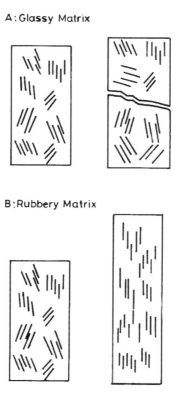

B : Rubbery Matrix

Fig. 30 A,B. Proposed model for the fracture of: **A** a glassy; **B** a rubbery polymer-clay exfoliated nanocomposite with increasing strain

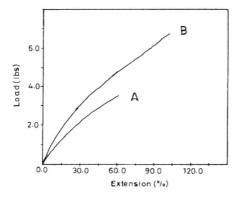

Fig. 31. Stress-strain curves for (A) a pristine PU elastomer and (B) a PU-clay nanocomposite

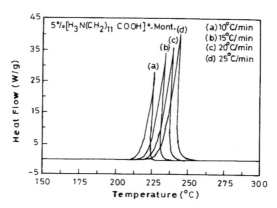

Fig. 32. DSC thermograms for epoxide polymerization catalyzed by 5 wt % $[NH_3(CH_2)_{11}COOH]^+$-MMT at heating rates of:10 °C/min, 15 °C/min, 20 °C/min, and 25 °C/min

6
Evaluation of Thermodynamic Parameters

Wang and Pinnavaia [28] determined several kinetic and thermodynamic parameters for their clay -polyether nanocomposite systems. Heats of reaction for $[H_3N(CH_2)_{11}COOH]^+$-MMT and $[H_3N(CH_2)_5COOH]^+$-MMT as a function of clay concentration decreased linearly with increasing clay concentration, implying that the heat of reaction was primarily due to epoxide polymerization.

Activation energies for a clay-polyether composite system were evaluated by obtaining the temperature dependence of epoxide polymerization (EPON-828) in the galleries of 5 wt % $[H_3N(CH_2)_{n-1}COOH]^+$-MMT (n=6, 12) from DSC.

Plots of DSC curves at various heating rates (Fig. 32) showed that the peak maximum temperature (T_p) increased with increasing heating rate (φ). Activation energy (E_a), was evaluated by using the empirically deduced equation [101–104] relating T_p and φ, regardless of the reaction order:

$$d \ln\varphi/d \ (1/T_p) \cong -1.052 \ E_a/R$$

and the preexponential factor (A) was determined from the expression [101–102, 105]

$$A \cong \varphi E_a \exp(E_a/RT_p)/RT_p^2$$

and $\Delta S^{0\#}$ was determined by the relation

$$\Delta S^{0\#} = R \ln \{Ah(C^0)^{m-1}/k \ T_e^m\}$$

where m is the molecularity of an elementary reaction or the number of molecules that come together to form the activated complex, $(C^0)^{m-1}$ is the factor re-

Table 8. Kinetic parameters for clay-polyether nanocomposites formed from $[H_3N(CH_2)_nCOOH]^+$-MMT-EPON-828 system

Interlayer cation	PDT^a (°C)	K_a (kJ/mol)	A (s$^{-1)}$	$\Delta S^{o\#}$ (eu)
$[H_3N(CH_2)_5COOH]^+$-MMT	248±1	105±1	3.06×10^8	−24.8±0.1−
$[H_3N(CH_2)_{11}COOH]^+$-MMT	229±1	108±1	1.80×10^9	21.2±0.1

^{a} Polymerization-delamination temperature determined at a heating rate of 20 °C/min and a clay loading of 0.5 wt %

quired to keep the equilibrium constant dimensionless, and h and k are the Planck and Boltzman constants, respectively. For bimolecular acid-catalyzed epoxide ring openings, m=2.

Table 8 compares [106,107] the various parameters thus evaluated for the formation of clay-polyether nanocomposites from $[H_3N(CH_2)_5COOH]^+$ and $[H_3N(CH_2)_{11}COOH]^+$ montmorillonite.

7
Conductivity Characteristics

7.1
PANI-Fluorohectorite Intercalates

Mehrotra and Giannelis [36, 37] prepared highly oriented multilayered PANI films by intercalative polymerization of ANI in a synthetic mica type silicate fluorohectorite. The insulating hybrid became conductive when exposed to HCl vapors with an in-plane conductivity of 0.05 S/cm. The conductivity of the multilayered films was highly anisotropic with an in-plane conductivity 10^5 times higher than the conductivity in the direction perpendicular to the film. This feature was consistent with increased charge localization and decreased conjugation length of the chains. The enforced conformation of the polymer as well as the existence of the insulating host layers disrupted the three-dimensional delocalization required for fully achieving the metallic state, which was accompanied by decreased conductivity.

7.2
PNVC-MMT Nanocomposite System

Table 9 represents the d.c. conductivity values of the PNVC-MMT composite [32, 40] along with the corresponding data for unmodified PNVC, MMT, and some PNVC based materials. Remarkably, the conductivity of the PNVC-MMT composite was enhanced 10^{10} times relative to that of unmodified PNVC. This feature is significant since conducting composites based on PNVC have been reported only in a few cases.

Table 9. D.C. conductivity data of PNVC-MMT composite and unmodified PNVC

Entry No.	Materials	Conductivity (S/cm)	Reference
1.	Unmodified PNVC	10^{-12} to 10^{-16}	32, 40
	Modified PNVC		
2.	PNVC-PPY blend	3.3×10^{-6}	108
3.	PNVC-PPY composite	5.0×10^{-2}	109
4.	PNVC-PPY (FeCl$_3$/Aq)	$(2.2$ to $3.5)\times10^{-3}$	110
5.	PNVC-Carbon black composite	1.0 to 1.2	111
6.	PNVC-MMT composite.	10^{-6}	32
7.	PNVC-FeCl$_3$ nanocomposite	3.1×10^{-5} [0.0074][a]	33
		5.2×10^{-5} [0.0289][a]	
8.	Montmorillonite (MMT)	3.4×10^{-7}	32

[a]Values in the parenthesis indicate FeCl$_3$ loading of MMT [ref. Table 2.]

7.3
Effect of FeCl$_3$ on Conductivity

The conductivity of the PNVC-MMT nanocomposite [33] prepared without FeCl$_3$ (entry 6, Table 9) was appreciably lower than the same made in the presence of FeCl$_3$ (entry 7, Table 9). Furthermore, a 3.4-fold increase in the FeCl$_3$ loading of MMT enhanced the conductivity of the composite about 1.7 fold. Thus, these trends in the conductivity data were consistent with the expected doping effect of FeCl$_3$. In a previous work on NVC-FeCl$_3$-toluene water system [110] it was observed that the conductivity of the PNVC continued to increase up to a certain FeCl$_3$/NVC weight ratio, and fall thereafter with further increase in the FeCl$_3$ concentration in the system. Unfortunately, the onset of gelation of PNVC at higher FeCl$_3$ impregnation level became a deterrent factor in establishing the definite trend in the variation of the conductivity of the composite with FeCl$_3$ loading.

7.4
PPY-MMT Nanocomposite System

The PPY-MMT (bulk) composite exhibited a d.c. conductivity (Table 3) of the order of 1.3×10^{-5} S/cm, which was comparable to the conductivity (2×10^{-5} S/cm) of the PPY-SiO$_2$ nanocomposite system reported by Armes et al. [50]. The composite obtained from PY-MMT (FeCl$_3$/bulk) system showed the expected enhancement in conductivity due to doping. For the composite obtained from the PY-MMT (FeCl$_3$/H$_2$O) [35] system, where the FeCl$_3$ impregnation level in MMT was varied twofold (entries 10–12, Table 3), the enhancement in the conductivity of the resultant PPY-MMT composite was about 1.8-fold.

7.5
PANI-MMT Nanocomposite System

Biswas and Sinha Ray [38] observed that the conductivity of the PANI-MMT nanocomposites prepared at fixed MMT and oxidant amounts in the initial feed increased steadily with increasing amount of ANI in the charge (entries 9–11, Table 4). This was apparently due to a parallel increase in the extent of conducting PANI loading per gram of the composite. The conductivities of the PANI-MMT composites increased with the amount of oxidant in the initial charge (entries 1,2, and 4, Table 4). In this case also, both the overall PANI yield as well as the PANI loading per gram of the composite increased with the amount of the oxidant. Hence, the observed conductivity trend was reasonable. The conductivities of the PANI-MMT composites increased with an increase of MMT amount in the initial feed, (entries 2, 6–8, Table 4), while both the total conversion to PANI as well as the amount of PANI loading per gram of composite decreased. Since the inherent conductivity of MMT was low, the above trend was puzzling, and the conductivities should have been expected to fall rather than increase, a trend parallel to PANI loading as in the other two instances referred to above. Armes et al. [49, 50, 55] pointed out that the electrical properties of the PANI oxide based composites were virtually the same as that of PANI-PVA or PANI-PVP composites of comparable PANI loading. Accordingly, the presence of the insulating silica or any other materials like MMT would not affect the nature of conduction.

It is well known [112] that the electrical conductivity of PANI can be widely varied by acidic protonation. With increase in the MMT amount, the conversion (Table 4) to PANI decreased. Since the proton concentration in the media was kept fixed (2 mol/l) in all the feed, the extent of protonation of PANI was expected to increase with decreasing PANI content in the system. Protonation in PANI would lead to the formation of radical cations by an internal redox reaction causing reorganization of electronic structure to produce semiquinone radical cations and accordingly the conductivity would be enhanced.

In general, at a mass loading of about 50% PANI, the volume fraction of the conducting PANI was well above the critical percolation threshold level of around 16% [113] so that conductivity as high as 1 S/cm should have been expected. The lower conductivity values ($\sim 10^{-3}$ S/cm) consistently observed in the PANI-MMT composites possibly resulted from the presence of low conducting MMT in the materials. In fact, entry 5, Table 4 shows a much higher conductivity for the PANI prepared under the same conditions without any MMT.

7.6
Conductivity Anisotropy Studies in MMT-Based Composites

Clay minerals such as Na-MMT exhibit appreciable ionic conductivities when swollen by water and polar polymers are known to mobilize Na^+ ions in Na-MMT [114–116].

NaMMT, a naturally occurring mineral from the clay group smectites, has a structure consisting of extended anionic layers balanced by mobile interlayer cations and a unit formula of $Na_{0.6}[(Mg_{0.6}Al_{3.4})Si_8O_{20}(OH)_4]$. Smectites are essentially polyelectrolytes with fixed anions. The identity of current carriers is less ambiguous in this type of electrolyte than in salt solutions, where both anions and cations are mobile. The charge sites in MMT are well separated, so ion pairing with the mobile cation is attenuated. To reduce further the attractive forces between the cation and the aluminosilicate sheets, a variety of solvating species such as PEO [117–121], MEEP [76], poly[oxymethylene oligo(ethylene oxide)] [122], cryptands [123], and crown ethers [123] were intercalated into the clays. The cationic mobility of these composite electrolytes was highly anisotropic, and greatly enhanced in comparison with the parent clay. For PEO-MMT composites, Ruiz-Hitzky and co-workers attributed the conductivity enhancement to increased layer separation and factors associated with relaxations of the polymer chain [123]. Lerner and Wu suggested that the polymer decreased the interactions between the cation and the negatively charged clay surface and thereby increased the conductivity [123]. Giannelis et al. [27] probed polymer dynamics and lithium ion transport in PEO-lithium-MMT composites with a variety of solid state NMR techniques (Fig. 33). At low temperatures, where the local segmental motion of the polymers was quiescent, a typical powder pattern was observed for both bulk d-PEO and the intercalated d-PEO. With an increase in temperature a central peak developed at lower temperature (250 K) in the intercalated d-PEO than in the bulk d-PEO (270 K). The breadth of the central peak from the intercalated d-PEO was substantially narrower than that from the bulk d-PEO. The central peak was believed to result from increased segmental motion causing temporal averaging of the signal.

Recently, Shriver and Ratner examined the role of intercalated MEEP-NaM-MT composite in long-range ion transport by an analysis of conductivity-tem-

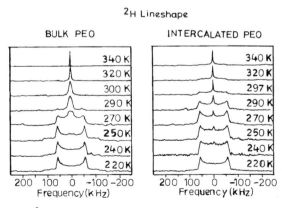

Fig. 33. Comparison of ^2H NMR line shapes for bulk d-PEO and Li-fluorohectorite-intercalated d-PEO as a function of temperature

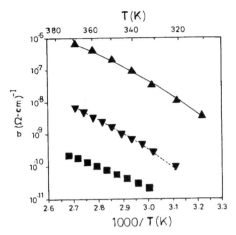

Fig. 34. Temperature dependent conductivities (σ) of MEEP-NaMMT parallel to the composite layers, ($\blacktriangle$) VTF fit; (-) (MEEP). NaMMT perpendicular to the composite layers, ($\blacktriangleleft$) VTF fit (–); Pristine Na.MMT ($\blacksquare$) measured perpendicular to the layers

Table 10. VTF Parameters for MEEP-Na MMT

Sample	$\sigma_0(\Omega^{-1}cm^{-1}\,K^{1/2})$	$\beta(J)$	$T_0\,(K)$
σ_{perp}	2.2×10^{-3}	1.2×10^4	218
σ_{para}	2.1×10^{-1}	1.3×10^4	204

perature studies [124]. Figure 34 suggests that the conductivity of the MEEP-NaMMT was substantially enhanced relative to pristine Na-MMT, with a $\sigma_{para}/\sigma_{perp}$ ratio of about 100. Both the parallel and perpendicular conductivities of the MEEP-NaMMT satisfied the Vogel-Tammann-Fulcher (VTF) equation commonly used to fit the conductivities of polymer electrolytes (Table 10) [125–132]:

$$\sigma = \sigma_0\, e^{-\beta/R(T-T_0)}$$

The VTF dependence was strongly suggestive of coupling between high-amplitude segmental motion and long-range cation transport in these composites. The conductivity anisotropy arose from the variation in the σ_0 term.

8
Water Dispersibility of MMT-Based Nanocomposites of PNVC, PPY, and PANI

Specialty polymers like PPY or PANI suffer from the obvious disadvantage of their general intractability which limits their practical application [133]. Stable aqueous dispersions of these polymers could be obtained by polymerizing the corresponding water soluble monomers in presence of nanodimensional inorganic oxides such as SiO_2 [49, 51], MnO_2 [52], ZrO_2 [53, 54] in water. Biswas and Sinha Ray also recently developed the conditions for preparing stable aqueous dispersions of PNVC [33].

However, for MMT-based nanocomposites of these polymers, achieving stable aqueous dispersions was also significant from the point of view of their practical applications. In this regard, Biswas and Sinha Ray [35, 38] examined several relevant conditions for obtaining stable aqueous dispersions (Table 11).

The use of a polymeric dispersant was essential for the production of stable dispersions of PPY-MMT and PANI-MMT nanocomposites. XRD analysis did not reveal any change in the d_{001} spacing of MMT after sonication in presence of the polymeric dispersant. So intercalation of PVP in the MMT layer and eventually trapping of the polymers to yield a stable dispersion was possibly not important. Stable aqueous dispersions of intractable PPY and PANI were successfully prepared with nanodimensional inorganic oxides such as SiO_2 [49, 97], MnO_2 [52], and ZrO_2 [53, 54] as particulate dispersants.

Previous studies on the synthesis of PANI dispersions reported a lack of success with PVA and other polymeric dispersants [133–135], the probable reason

Table 11. Dispersibility of MMT-based nanocomposites of PANI, PPY and PNVC in aqueous medium

Entry No.	Condition applied for obtaining aqueous dispersion	Observations
1a	PANI-MMT composite was added to water and the mixture was sonicated for 5 h	Dispersion stable up to three days
b	PANI-MMT composite was added to an aqueous solution of PVP and the system was sonicated for 5 h	Stable colloid but not redispersible
c	MMT was added to an aqueous solution of PVP which was stirred for 2 h; the $(NH_4)_2S_2O_8$ solution and ANI were added serially and the mixture was stirred magnetically for 20 h	Permanently stable dispersion and redispersible
2a	PPY-MMT composite was added to water and the mixture was sonicated for 5 h	Dispersion stable up to 40 h
b	PPY-MMT composite was added to an aqueous solution of PVP and the mixture was sonicated for 5 h	Stable colloids but not redispersible
c	MMT was added to an aqueous solution of PVP which was stirred for 2 h, the $FeCl_3$ solution and PY were added serially and the mixture was stirred magnetically for 16 h	Permanently stable colloid and redispersible
3a	PNVC-MMT composite was added to water and sonicated for 5 h	Unstable dispersion
b	MMT was added to PVP solution in alcohol and the system was stirred for 2 h. A solution of PNVC in benzene was added and the system was stirred magnetically for 22 h	No composite formation

being the poor adsorption of these stabilizers on the PANI surface. On the contrary, Gospodinova et al. [136] claimed successful dispersions using PVA due to efficient grafting of PVA onto the PANI particles. To overcome the problem of low adsorption of conventional stabilizers, Armes et al. [137–139] used tailor-made stabilizers, which could undergo graft copolymerization with ANI and thus get anchored on the PANI particles to assist stable dispersion.

Table 11 shows that PVP was an effective stabilizer if used with MMT prior to addition of the polymerizing monomers (ANI and PY) and the oxidants. Apparently, this procedure would help surface adsorption of the available polymeric stabilizer particles by the precipitating polymers.

The observed failure of the PNVC-MMT nanocomposite system to yield a stable dispersion in the presence of PVP in contrast to PPY/PANI based systems is indeed interesting. No convincing explanation for this difference is available as of now, but water compatibility of the base monomers (PY or ANI water-soluble but NVC insoluble) may be an important factor. In addition, since PNVC is bulkier than PPY or PANI, steric factors may limit the stabilization of the PNVC moieties by PVP relative to that for PPY or PANI

9
Conclusion: Status and Prospects of Application of Polymer-Clay Nanocomposites

The PU-clay nanocomposite was identified to be a material of high optical transparency, which make them especially attractive for high performance packaging materials, protective films, and high barrier sealants [29].

9.1
Advantages of PLS Nanocomposites

PLS nanocomposites possess several advantages:
1. They are lighter in weight compared to the conventionally filled polymers because high degree of stiffness and strength are realized with far less high density inorganic materials.
2. Their mechanical properties are potentially superior to fiber reinforced polymers because reinforcement from the inorganic layer will occur in two-dimensions rather than one-dimension without special effects to laminate the composites.
3. They exhibit outstanding diffusional properties. PLS nanocomposites are unique model systems in which to study the states and dynamics of polymers in confined environments, which are fundamental aspects of many industrially important fields as tribology, adhesion, fabrication, and catalysis [87].

Solvent-free electrolytes are of much interest because of their charge-transport mechanisms and their possible applications in electrochemical devices. In this regard, intercalates of MMT with PEO [117–121], MEEP [76], poly[oxymethylene oligo(ethylene oxide)] [122] cryptands, [122] and crown ethers [122] can be developed as potential materials. The cationic mobility of these composites is highly anisotropic and greatly enhanced in comparison with the parent clay.

9.2
Specialty Polymer- Based Nanocomposites

Probable applications in various fields like electrochromic displays, electronic devices, modified electrodes, chemical- and bio-sensors etc. [133] may be envisaged. Potentially, nanodimensional oxide- (SiO_2, MnO_2 and ZrO_2) based composites of these polymers have shown promise as optical materials [50]. Conducting polymer-coated low T_g latexes of sub-micron (50–500 nm) dimensions are reported to find use in antistatic and anticorrosion applications [113]. In the light of this information it would be of interest to explore the usability of the corresponding MMT-based composites in these directions.

Acknowledgments. The authors thank CSIR, New Delhi, for generous funding of an Emeritus Scientist Scheme in favor of MB, Prof. S, Ghosh, Head of the Department of Chemistry and Prof. N.C. Mukherjee, Principal, Presidency College, Calcutta for extending all facilities.

References

1. Gaylord NG, Encer.H, Davis.L, Takahashi.A (1980) In: Carraher CE, Tsuda M (eds) Modification of polymers. ACS Symp Ser 121, Washington, DC
2. Voet A (1980) J Polym Sci Macromol Rev 15:327
3. Theng BKG (1974) The chemistry of clay-organic reactions. Wiley, New York
4. Theng BKG (1979) Formation and properties of clay polymer complexes. Elsevier, New York
5. Giannelis EP, Mehrotra V, Tse OK, Vaia RA, Sung TC (1992) In: Rhine WE, Shaw MT, Gottshall RJ, Chen Y (eds) Synthesis and processing of ceramics: scientific issues. MRS Proc. Pittsburg PA, p 249:547; Giannelis EP (1996) Adv Mater 8:29
6. Lan T, Pinnavaia TJ (1994) Chem Mater 6:2216
7. Sugahara Y, Sugiyama T, Nagayama T, Kuroda K, Kato C (1992) J Ceram Soc Jpn 100:413
8. Vaia RA, Ishii H, Giannelis EP (1993) Chem Mater 5:1694
9. Blumstein A (1965) J Polym Sci Part A Polym Chem 3:2653; Blumstein A, Malhotra SL, Watterson AC (1970) J Polym Sci Part A Polym Chem 8:1599
10. Blumstein R, Blumstein A, Parikh KK (1994) Appl Polym Symp 25:81
11. Sugahara Y, Satakawa S, Kuroda K, Kato C (1988) Clays and Clay Miner 36:343
12. Bergaya F, Kooli F (1991) Clay Miner 26:33
13. Kato C, Kuroda K, Takahara H (1981) Clays and Clay Miner 29:294
14. Friedlander HZ (1964) Chem Eng News 42:42
15. Friedlander HZ, Frink CR (1964) J Polym Sci B 2:475
16. Wheeler A (1958) US Pat 2,847,391
17. Greenland DJ (1963) J Colloid Sci 18:647
18. Francis CW (1973) Soil Sci 115:40
19. Parfilt RL, Greenland DJ (1970) Clay Minerals 8:305
20. Schamp N, Huylebroeck (1973) J Polym Sci, Polym Symp 42:553
21. Okada A, Kawasumi M, Vsuki A, Kojima Y, Kurauchi T, Kumigaito O (1990) Mater Res Soc Symp Proc. 171:45

22. Okada A, Kawasumi M, Kurauchi T, Kumigaito O (1987) Polym Prepr 28:447
23. Burnside SD, Giannelis EP (1995) Chem Mater 7:1597
24. Kojima Y, Fukumori K, Usuki A, Okada A, Kurauchi T (1993) J Mater Sci Lett 12:889
25. Yano K, Usuki A, Okada A, Kurauchi T, Kamigaito O (1993) J Polym Sci Part A: Polym Chem 31:2493
26. Kijima Y, Usuki A, Kawasumi M, Okada A, Kuraurchi T, Kamigaito O (1993) J Polym Sci Part A: Polym Chem 31:983; Mesersmith PB Giannelis EP (1993) Chem Mater 5:1064
27. Wong S, Vasudevan S, Vaia R, Giannelis EP, Zax D (1995) J Am Chem Soc 117:7568
28. Wang MS, Pinnavaia TJ (1994) Chem Mater 6:468
29. Messersmith PB, Giannelis EP (1994) Chem Mater 6:1719
30. Woods G (1990) The ICI polyurethanes book. Wiley, New York
31. Wang Z, Pinnavaia TJ (1998) Chem Mater 10:3769
32. Biswas M, Sinha Ray S (1998) Polymer 39:6423
33. Sinha Ray S, Biswas M (1999) J Appl Polym Sci 73:2971
34. Okari OC, Lerner M (1995) Mat Res Bull 30:723
35. Sinha Ray S, Biswas M (1999) Mat Res Bull 34:8
36. Mehrotra V, Giannelis EP (1991) Solid State Comm 77:155
37. Mehrotra V, Giannelis EP (1992) Solid State Comm 51:155
38. Biswas M, Sinha Ray S (2000) J Appl Polym Sci 77:2948
39. Pennwell RC, Ganguly BN, Smith TW (1973) J Polym Sci Macromol Rev 13:63 (and reference cited therein)
40. Biswas M, Das SK (1982) Polymer 23:1706
41. Biswas M, Mitra P (1991) J Appl Polym Sci 42:1989
42. Biswas M, Majumder A (1991) J Appl Polym Sci 41:1489
43. Mitra M, Biswas M (1992) J Appl Polym Sci 45:1685
44. Maeda S, Armes SP (1994) Mater Chem 4:935
45. Markham G, Obey TM, Vincent B (1990) Colloids Surf 51:239
46. Terrill NJ, Crowley T, Gill M, Armes SP (1993) Langmuir 9:2093
47. Brus LE, Brown WL, Andres RP, Averback RS, Goddard WA, Kaldor A, Louie SC, Moskovits M, Peercy PS, Riley SJ, Siegel RW, Spaepen FA, Wang Y (1989) J Mater Res 4:704
48. Appenzeller T (1991) Science 254:1300
49. Stejskal J, Kratochvil P, Armes SP, Lascelles SF, Riede A, Helmstedt M, Prokes J, Krivka I (1996) Macromolecules 29:6814
50. Armes SP, Gottesfeld S, Beery JG, Garzone F, Agnew SF (1992) Polymer 32:2325
51. Sinha Ray S, Biswas M (1998) Mat Res Bull 33:533
52. Biswas M, Sinha Ray S, Liu YP (1999) Synthetic Metals 105:99
53. Sinha Ray S, Biswas M (2000) Synthetic Metals 108:231
54. Bhattacharya A, Ganguly KM, De A, Sarkar S (1996) Mat Res Bull 31:527
55. Maeda S, Armes SP (1995) Chem Mater 7:171
56. Biswas M, Maity NC (1979) Adv Polym Sci 31:47
57. Biswas M, Maity NC (1982) J Macromol Sci Chem A18(3):477
58. Biswas M, Maity NC (1980) Polymer 21:1344
59. Akelah A, Moet A (1994) J Appl Polym Sci, Appl Polym Symp 55:153
60. Biswas M, Roy A (1995) Eur Polym J 31:725
61. Biswas M, Roy A (1996) J Appl Polym Sci 60:43
62. Moet A, Akelah A (1993) Mater Lett. 18:97
63. Moet A, Akelah A, Hiltner A, Baer E (1994) Proc.Mater Res Soc San Fancisco April 2–8
64. Akelah A, Salahuddin N, Hiltner A, Baer E, Moet A (1994) Nanostructured Mater 4:3
65. Pinnavaia TJ (1983) Science 220:365
66. Solomon DH, Hawthorne DG (1991) Chemistry of pigments and fillers. Krieger, Malabar, FL
67. Whittinggham MS, Jacobson AJ (eds) (1982) Intercalation chemistry. Academic Press, New York

68. Cao G, Mallouk TE (1991) J Solid State Chem 94:59
69. Aranda P, Ruiz-Hitzky E (1992) Chem Mater 4:1395
70. Fukushima Y, Inagaki S (1987) J Incl.Phenom. 5:473
71. Usuki A, Kojima Y, Kawasami M, Okada A, Kurauchi T, Kamigaito O (1990) Poly Prepr ACS Div Polym 31(2):651
72. Usuki A, Mizutani T, Fukushima Y, Fujimoto M, Fukumori K, Kojima Y, Sato N, Kurauchi T, Kamigaito O (1989) US Pat 4,889,885
73. Fujiwara S, Sakamota T (1976) Japan Pat 51:109,998
74. May CA (ed) (1988) Epoxy resin, 2nd edn. Marcel Dekker, New York
75. Lee H, Neville K (1967) Handbook of epoxy resins. McGraw-Hill, New York
76. Hutchison JC, Bissessur R, Shriver DF (1996) Chem Mater 8:1597
77. DeArmit C, Armes SP (1991) J Colloids Interface Sci 150:134
78. Grim RE (1953) Clay minerology. McGraw-Hill, New York
79. Pillon JE, Thompson ME (1991) Chem Mater 3:377
80. Kantzidis MG, Wu C (1989) J Am Chem Soc 111:4139
81. Kanatzidis MG, Wu CG, Marcy HO, Degroot DC, Kannewurf CR (1990) Chem Mater 2:222
82. Liu YJ, Degroot DC, Schindler JL, Kannewurf CR, Kanatzidis MG (1991) Chem Mater 3:992
83. Kanatzidis MG, Wu CG, Degroot DC, Schindler JL, Benz M, Legoff E, Kannewurf CR (1993) In: Fisher J (ed) NATO ASI, Chemical Physics Intercalation II; Plenum Press, New York; Kanotzidis MG, Marcy HO, McCarthy WJ, Kannewurf CR, Marks TJ (1989) Solid State Ionics 32/33:594; Kanotzidis MG, Marcy HO, McCarthy WJ, Kannewurf CR, Marks TJ (1989) J Am Chem Soc 111:4139
84. Diviglpitiya WMR, Frindt RF, Morrison SR (1991) J Mater Res 6:1103
85. Tanaka Y, Bauer RS (1988) In: May CA (ed)Epoxy resins. Marcel Dekker, New York
86. Tanzer W, Reinhardt S, Fedtke M (1993) Polymer 34:3520
87. Tanaka Y, Kakiuchi H (1963) J Appl Polym Sci 7:1063
88. Khrishnamoorti R, Vaia RA, Giannelis EP (1996) Chem Mater 8:1728
89. Sauer BB, Haiao BS (1993) J Polym Sci Polym Phys Edn 31:917
90. Sauer BB, Dipaolo NV, Avakian P, Kampert WG, Starkweather HW (1993) J Polym Sci Polym Phys Edn 31:1851
91. Sauer BB, Beckerbauer R, Wang l (1993) J Polym Sci Polym Phys Edn 31:1861
92. Sauer BB, Avakian P (1992) Polymer 33:5128
93. Keddie JL, Jones RAL, Cory RA (1994) Europhys. Lett. 27:59
94. Keddie JL, Jones RAL, Cory RA (1994) Faraday Discuss 98:219
95. Wallace WE, VanZanten JH, Wu W (1995) Phys.Rev E52:R3329
96. Butterworth MD, Corradi R, JoHal J, Lascelles SF, Maeda S, Armes SP (1995) J Colloid Interface Sci 174:517
97. Maeda S, Armes SP (1995) Synth Met 73:151
98. Landry CJT et al. (1993) Macromolecules 26:3702
99. Huang HH, Wilkes GL, Carlson JG (1989) Polymer 30:2001
100. Huang HH, Orler B, Wilkes GL (1987) Macromolecules 20:1322
101. Prime RB (1981) In: Turi E (ed) Thermal characterization of polymeric materials. Academic, New York
102. Ozawa TJ (1970) Therm Anal 2:301
103. Doyle CD (1951) Anal Chem 33:79
104. Prime RB (1973) Polym Eng Sci 13:365
105. Peyer P, Bascom WD (1975) J Polym Sci, Polym Phys Edn 18:129
106. Swarim SJ, Wims AM (1976) Anal Calorim 4:155
107. Cizemciogly M, Gupta A (1962) SAMPLE Q April 16
108. Tetsuyoshi S, Kazumi H, Kaji N, Senehiro F (1987) Japan Pat JP 1987 62,108,400
109. Geissler U, Hallensleben ML, Toppave L (1991) Synth Met 40:239
110. Biswas M, Roy A (1994) Polymer 35:4470

111. Biswas M, Roy A (1993) Polymer 34:2903
112. Chiang JC, McDiarmid AG (1986) Synth.Met. 13:193; Zuo F, Angelopoulos M, MacDiarmid AG, Epstein AJ (1987) Phys Rev B36:3475
113. Prokes J, Krivka J, Stejskal J, Kratochvil P (1996) Abs: Partnership in Polymers, The Cambridge Polymer Conference, 30 Sept–2 Oct, p 100; Balberg L, Binenbaum N (1987) Phys Rev B 35:8749
114. Wang WL, Lin FL (1990) Solid State Ionics 40/41:125
115. Fan YQ (1988) Solid State Ionics 28:1596
116. Salde R, Braker J, Hiris P (1987) Solid State Ionics 24:289
117. Ruiz-Hitzky E, Aranda P, Casal B, Galvan (1995) J Adv Mater 7:154
118. Vaia RA, Vasudevan S, Wlodzimierz K, Scanlon L, Giannelis EP (1995) Adv Mater 7:154
119. Aranda P (1993) Adv Mater 5:334
120. Aranda P, Ruiz-Hitzky E (1992) ElectroChem Acta 37:1573
121. Ruiz-Hitzky E, Aranda P (1990) Adv Mater 2:545
122. Ruiz-Hitzky E, Casal B (1978) Nature 276:596
123. Wu J, Lerner M (1993) Chem Mater 5:835
124. Ratner MA, Shriver DF (1988) Chem Rev 88:109; Allock HR, Austin PE, Neenan TX, Sisko JT, Blonsky PM, Shriver DF (1986) Macromolecules 19:1508
125. MacCallum JR, Vincent CA (1987, 1989) Polymer Electrolytes Reviews. Elsevier, London, vols 1 and 2
126. Tonge JS, Shriver DF (1989) In: Lai JH (ed) Polymers for electonic applications. CRC Press, Boca Raton, FL, pp 194–200
127. Cheradame H (1982) In: Beonit H, Rempp P (eds) IUPAC Macromolecules. Pergamon Press, New York, p.251
128. Killis A, LeNest JF, Cheradame HJ (1980) J Polym Sci Macromol Chem Rapid Commun 1:595
129. Killis A, LeNest JF, Cheradame HJ, Cohen-Addad JP (1984) Solid State Ionics 14 :231
130. Druger SD, Ratner MA, Nitzan A (1985) Phys. Rev B 31:3939
131. Tipton AL, Lonergan MC, Shriver DF (1994) J Phys Chem 98:4148
132. Lonergan MC, Ratner MA, Shriver DF (1995) J Am Chem Soc 117:2344
133. Bhattacharya A, De A (1996) Progress in Solid State Chem 24:141
134. Banerjee P, Bhattacharyya SN, Mandal BM (1995) Langmuir 11:2414
135. Armes SP, Aldissi M (1989) J Chem Soc, Chem Comm 88
136. Gospodinova N, Mokreva P, Terlemezyan I (1992) J Chem Soc, Chem Comm 923
137. Tadors P, Armes SP, Luk SY (1992) J Mater Chem 125:2
138. Armes SP, Aldissi M, Gottesfeld M, Agnew SF (1990) Langmuir 6:1745
139. Simmons MR, Chaloner PA, Armes SP, Greaves SJ, Watts JF (1998) Langmuir 14:611

Accepted by: Prof. Suter / Dr. Osman
Received: May 2000

Author Index Volumes 101–155

Author Index Volumes 1–100 see Volume 100

Subject Index

Printing (Computer to Film): Saladruck, Berlin
Binding: Stürtz AG, Würzburg

T